Brickwork
and
Blockwork

Questions and Answers books are available on the following subjects:

QUESTIONS & ANSWERS

Brickwork
and
Blockwork

R. A. Daniel

Newnes Technical Books

THE BUTTERWORTH GROUP

UNITED KINGDOM
Butterworth & Co (Publishers) Ltd
London: 88 Kingsway, WC2B 6AB

AUSTRALIA
Butterworths Pty Ltd
Sydney: 586 Pacific Highway, Chatswood, NSW 2067
Also at Melbourne, Brisbane, Adelaide and Perth

CANADA
Butterworth & Co (Canada) Ltd
Toronto: 2265 Midland Avenue, Scarborough, Ontario M1P 4S1

NEW ZEALAND
Butterworths of New Zealand Ltd
Wellington: 26—28 Waring Taylor Street, 1

SOUTH AFRICA
Butterworth & Co (South Africa) (Pty) Ltd
Durban: 152—154 Gale Street

USA
Butterworth (Publishers) Inc
Boston: 19 Cummings Park, Woburn, Mass. 01801

First published 1977 by Newnes Technical Books,
a Butterworth imprint

© Butterworth & Co (Publishers) Ltd, 1977

ISBN 0 408 00263 8

Typeset by Butterworths Litho Preparation Department
Printed in England by Cox & Wyman Ltd., London Fakenham and
Reading

CONTENTS

PREFACE

The origins of brickwork are lost in antiquity. Young people have been learning the craft for several thousand years and helping to provide man with shelter — a basic necessity of life.

Modern times have in many ways simplified brickwork and in many cases changed its role from a structural to a cladding function. However, brickwork is still the major construction material for most work, in face of competition from all other building systems. The extensive use of blockwork in its own right and in conjunction with brick has probably been the greatest change which has taken place in the trade this century. The variety of blocks available and the methods of use have faced the bricklayer with the need to extend his technology and develop new, if similar skills.

Today's apprentice has only three years in which to learn his trade against five years for his predecessors. This drastic reduction in time has been brought about by starting the apprentice on a full time training course in a works training school or technical college, of six months' duration, followed by a day release course until the end of his apprenticeship.

This book aims to cover topics which appear in the C.I.T.B. new entrant training course and the C.G.L.I. Basic Craft Course. Most of the topics are a personal selection and have been dealt with as basic knowledge, necessary for these courses. The difficulty has been deciding what to leave out. Certain topics have been omitted purposely as they are dealt with in other books in this series.

R.A.D.

1

TOOLS

What are the basic measuring tools?

The four-fold boxwood rule of 1 m length and the 3 m steel tape are generally used for measurement of small units. The plastics faced linen or the steel tape of 30 m length are used to measure wall lengths and for setting out houses and other small buildings.

What is the builder's setting out square?

A wooden square with sides about 1 m length is used to set out square corners and angles (*Figure 1*).

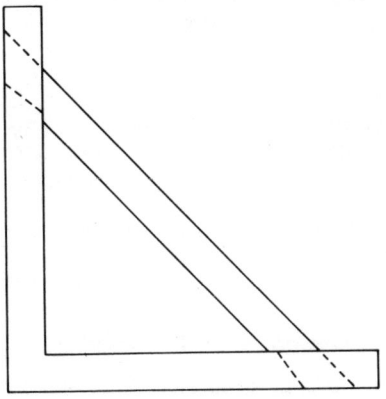

Figure 1. Wooden setting out square

How is a spirit level used?

The spirit level is a parallel piece of wood or aluminium 1 m length × 75 mm × 25 mm (*Figure 2*). The glass or clear

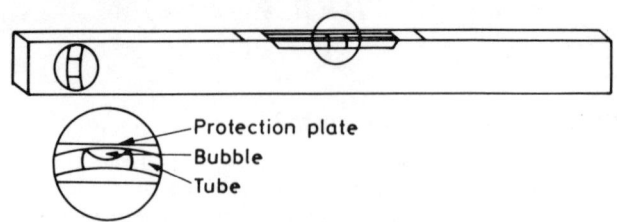

Figure 2. Spirit level

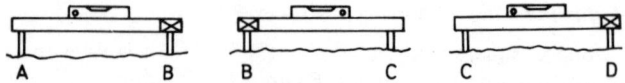

Figure 3. Reversing the level and straight edge when transferring level points

plastics tube has an air bubble in a coloured spirit. The tube is slightly curved to ensure that when correctly set in a level position the bubble will be central between two marks on the tube. The level is used with a straight edge and a series of intermediate points set up until the object point has been reached; level and straight edge are reversed each time to eliminate possible errors (*Figure 3*).

What other instruments are used for levelling?

The water level (*Figure 4*) is used to transfer and check levels where intervening obstacles prevent more direct levelling.

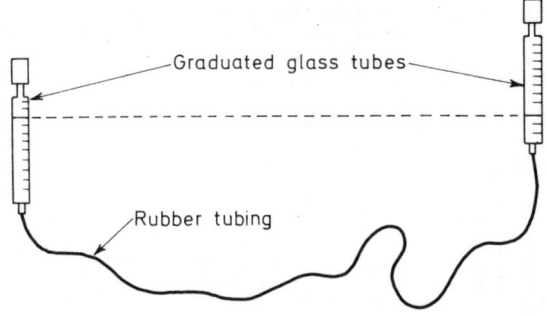

Figure 4. Water level

What are boning rods?

They are used for levelling from two fixed points (*Figure 5*). The rods are T-shaped and of the same size. Levelling is carried out by eyesight along the top edge of the rods.

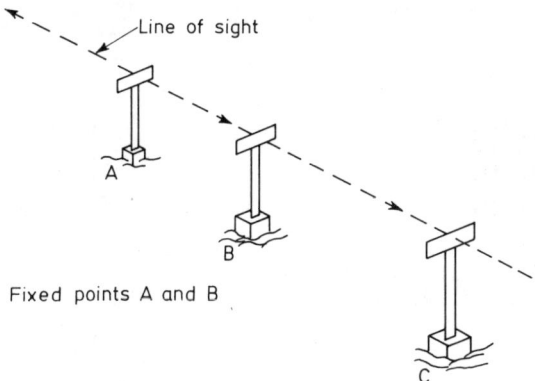

Figure 5. Boning rods

What tools are used to keep brickwork vertical?

Plumb rule and bob (*Figure 6*). The line will always hang in a vertical position and the centre line on the rule can be aligned with it to give a truly plumb edge. The centre line

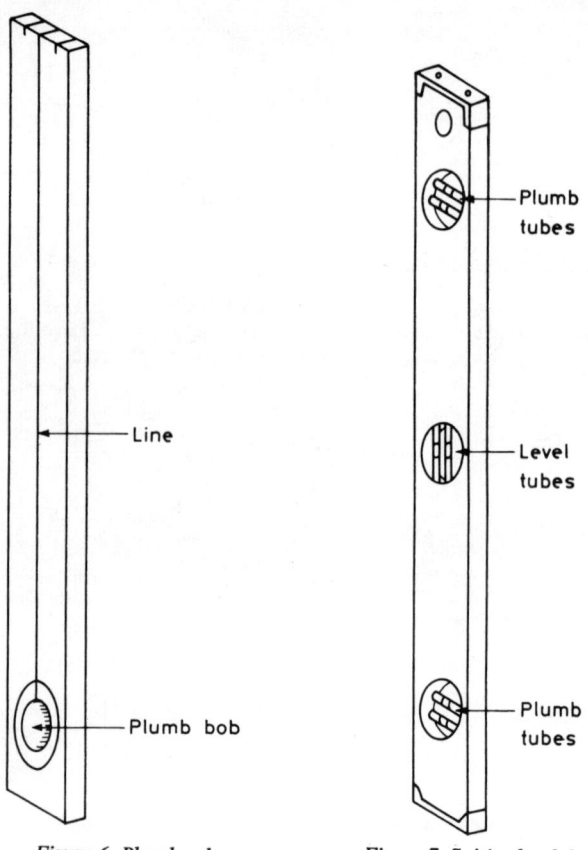

Figure 6. Plumb rule

Figure 7. Spirit plumb level

4

must be parallel with the edge of the rule to ensure accuracy. The spirit plumb (*Figure 7*) is very similar to the level except that the bubble tube is set at right-angles to the edge of the rule. When the bubble is between the two marks, the edge of the rule is truly vertical.

How can spirit level tubes be checked for accuracy?

The level is set on two screws set in a firm wood base and 150 mm less than the length of the level apart. One screw is

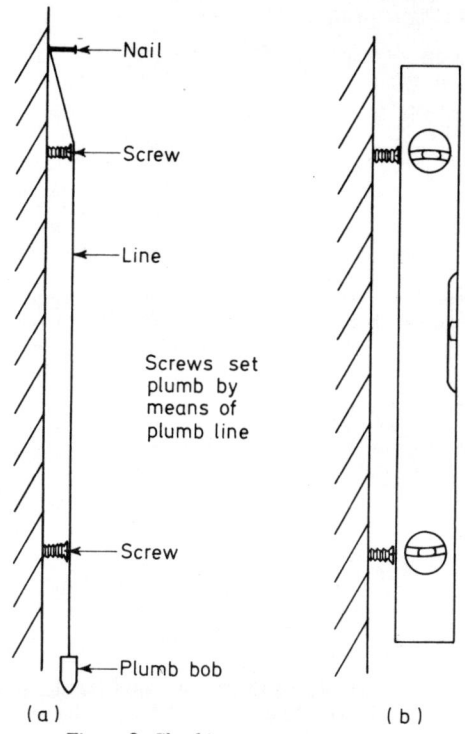

Figure 8. Checking a spirit plumb rule

adjusted until the bubble comes to rest between the two lines on the tube. The level is now picked up and reversed. If the bubble again comes to rest between the two lines then the level is correct. In a similar way, the spirit plumb can be checked for vertical by reversing on two screws set in a vertical plane. If the level is found to be inaccurate and is an adjustable type, adjustments can be made and checked in the same way (*Figure 8*).

What is a laying trowel?

The tool used to pick up and spread a mortar bed and to place the cross joints on the bricks. The trowel is used to tap the bricks down on to the bed and can be used for rough cutting of soft bricks.

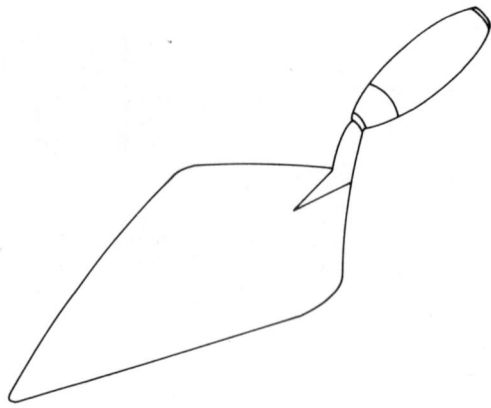

Figure 9. Laying trowel

A trowel blade is 225–325 mm in length and 75–150 mm in width. A solid forged tang passes through the wood handle to a ferrule which protects the end (*Figure 9*).

6

How is the laying trowel used?

A suitable quantity of mortar is cut from the heap on the board and pushed to one side to ensure it is cohesive (*Figure 10(a)*). The trowel is then slid under and the mortar picked up. The trowelfull is spread in position by drawing the arm sharply backward, at the same time twisting the wrist and allowing the mortar to leave the trowel.

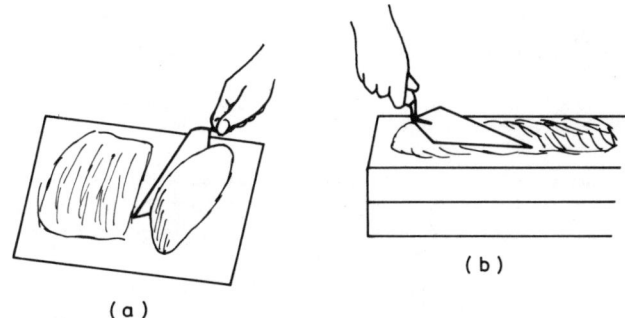

(b)

(a)

Figure 10. Picking up and spreading mortar

The point of the trowel is then drawn through the mortar to ensure even bedding (*Figure 10(b)*). Cross joints are placed on the end of a brick by taking a small quantity of mortar on the trowel and applying it to the brick, striking the trowel off the edge of the brick to ensure firm adhesion and a full flush joint. The brick is then laid in position.

What tools are used to cut bricks to length?

Rough cutting is done with the cutting edge of a brick hammer (*Figure 11*). More accurate work is done using a bolster chisel and lump hammer (*Figure 12*).

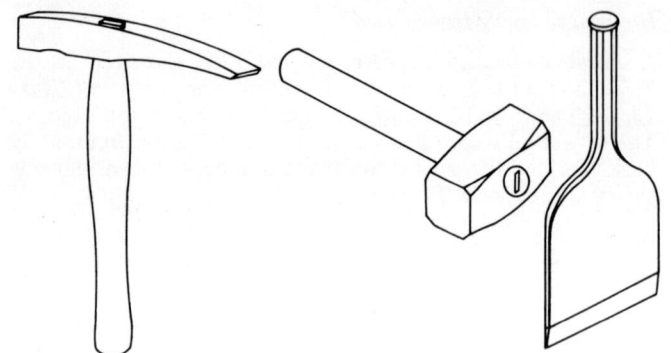

Figure 11. A brick hammer

Figure 12. Lump hammer and bolster chisel

The cut should be measured and marked with a pencil, then marked in the brick, using the bolster chisel and a light blow of the hammer. The bolster chisel is then placed on the mark

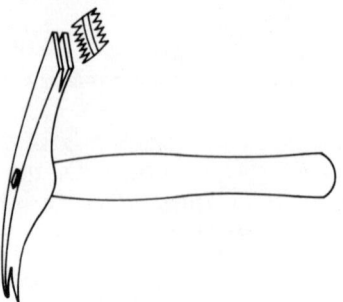

Figure 13. Double-end comb hammer or skutch

on the face side and struck a hard blow to cut the brick. The comb hammer (*Figure 13*) is used to remove surplus material after a bolster cut where greater accuracy is needed.

Which tools are used to give a neat finish to the mortar joints?

The pointing trowel is a smaller, lighter version of the laying trowel, being 150 mm long and 50 mm wide. It is used to form weathered joints by drawing the trowel along the bed joint at an angle (*Figure 14(a)*) to improve the weather resistant qualities of the wall. When used the opposite way, a struck joint is formed (*Figure 14(b)*).

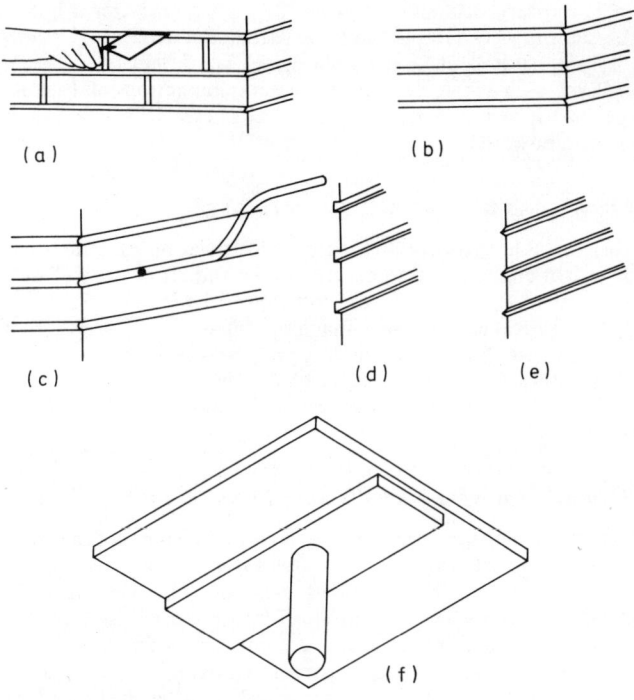

Figure 14. (a) Weather joint. (b) Struck joint. (c) Round recessed joint. (d) Square joint. (e) V-joint. (f) Wooden hand hawk

9

Recess jointing effects can be obtained by using metal jointers such as the round (*Figure 14(c)*), square (*Figure 14(d)*), or V (*Figure 14(e)*). The mortar used for pointing is held on a wooden hand hawk 150 mm square (*Figure 14(f)*).

What portable power tools are used to drill holes in brickwork?

Rotary and percussion type drills are used to drill holes in brick and blockwork, for fixings, cables and pipes.

The rotary drill simply turns the bit which bores a hole in the masonry by scraping action. Percussion drills also turn the bit, but hammer it at the same time. The effect is to use the drill bit as a chisel as well as the scraping action, and greatly speeds up the drilling of the hole. These drills are usually electric powered.

Which power tools are suitable for cutting?

The portable carborundum disc saw, driven by electric motor or petrol engine, is used to cut bricks and stones as well as all kinds of sheet materials, concrete, pipes and tiles.

The portable hand-held machines are dry cutters and create a lot of dust. The fixed bench type cutters incorporate a pump to spray water on to the cutter blade.

In both cases, protective clothing and eyeshields must be worn.

What other jobs are generally done by power tools?

Chase cutting, the forming of grooves or channels for electric conduits, water heating pipes and gas pipes, also for sinking switch boxes and similar items into the wall. This operation is carried out by fitting a chasing bit into a percussion tool for hard, dense walling, or using a comb chisel for low density materials such as lightweight partition blocks.

The disc grinder is a high speed abrasive disc used for rubbing down concrete to expose the aggregate and general grinding of rough cut brick surfaces.

Is protective clothing necessary when using power tools?

Drilling and cutting work with power tools in brick, concrete and similar materials creates a lot of dust and small chippings. It is essential to protect the eyes with an industrial eyeshield, though a full face shield gives better protection when using an abrasive disc cutter. Leather or p.v.c. gloves are worn to protect the hands, but thick heavy type gloves should not be worn as they prevent the operator getting a firm grip on the handle.

Safety boots should be worn as a protection to the feet when using a cutting tool in a vertical position to prevent injured toes if the tool slips.

What safety checks should be carried out before using electric power tools?

The voltage of the motor in a power tool must be checked to ensure that it corresponds with the power supply. Voltage is marked on a plate attached to the tool, and in construction work is 110 V. A transformer is used to reduce the standard 240 V supply. Non-standard plugs are fitted to these tools to make sure they cannot be plugged into a standard 240 V socket. The lead cable should be rubber or p.v.c. covered, with no cuts or damage to it; rubber sheathed plugs and connections to extension leads prevent water penetration in the event of the cable lying on a wet floor or being used outside in the rain. The handgrips should not be damaged and in many cases may have to be adjusted for left- or right-handed people. Another source of danger can be the chuck or the tool holder — a check should be made that these function properly and clamp the bit or the tool securely. With a saw or grinder, a check should be made to ensure that the blade is the correct one for the job to be done and that the guards work properly.

What safety checks should be carried out on compressed air and petrol engined power tools?

These tools are not much used in brickwork and blockwork and the safety checks are similar to those given for electric power tools. Compressed air tools should be checked to see that the output of the compressor in N/mm^2 corresponds with the tool being used.

The air pipe from the compressor to the tool must be undamaged and correctly connected. Petrol engined power tools must not be used in a confined room as the fumes from the engine will overcome the operator; good ventilation is essential. Care must be taken when filling with petrol to prevent explosion and spillage. Both types of power tool need oil levels checked before use and at regular intervals, as manufacturer's instructions.

2

MATERIALS

What is a brick?

A rectangular block of hard, durable inert material, of a size
suitable to be handled with one hand, of standard dimensions
215 × 102.5 × 65 mm.

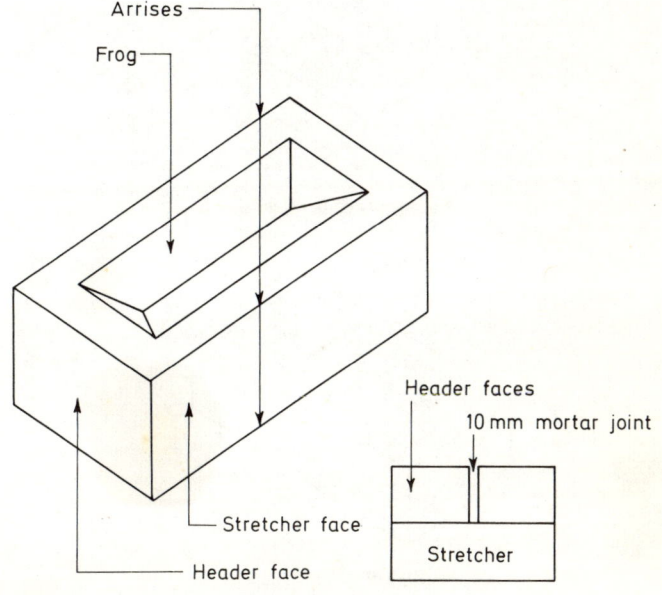

Figure 15. A brick

The long sideface of a brick is called the *stretcher face*. The end of the brick is the *header face*. The edges are known as the *arrises*. In the top of a pressed brick there is usually an indent known as a *frog*, on which the maker's name is generally stamped.

Two header faces and a 10 mm mortar joint equal one stretcher face in length (*Figure 15*).

What are bricks made from?

Clay bricks are fired bricks, earth or shale. They are formed by pressing in moulds, or by an extrusion and wire cutting process, then dried and fired in a kiln.

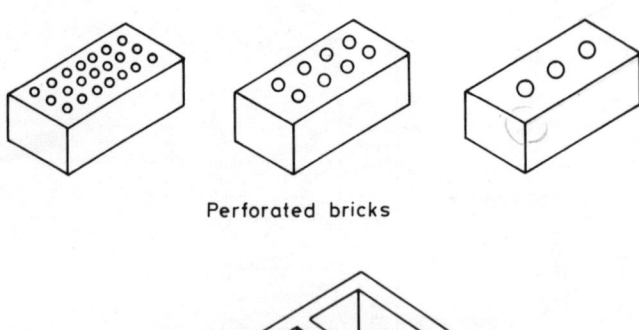

Perforated bricks

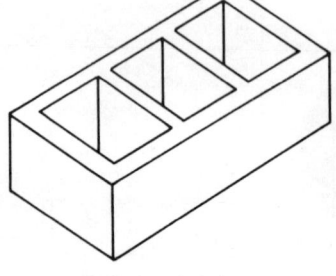

Cellular brick

Figure 16. Perforated bricks

14

Sand lime bricks are a mix of sand and hydrated lime, pressed and then cured in a high pressure steam autoclave.

Concrete bricks are a mix of aggregates and cement pressed or vibrated in moulds and steam cured.

Clay and sand lime bricks sometimes have holes pushed through them after the pressing or wire cut process. These are known as *perforated bricks* (*Figure 16*).

What are the three basic types of brick?

Common bricks are general purpose bricks which do not have a special face treatment. They are usually rough finish, provide a good key for plaster, do not have any particular colour or handling and are mainly used on work which is unseen or whose appearance is relatively unimportant.

Facing bricks have special care taken in manufacture and handling. Colour, surface texture, uniformity of shape and size are features of this type of brick.

Engineering bricks are manufactured so as to have very high crushing strength and very low water absorption. BS 3291 specifies two classes:

	Class A	*Class B*
Average compressive strength (MN/m^2) not less than:	69.0	48.5
Average absorption boiling or vacuum (% weight) not greater than:	4.5	7.0

What is the difference between a pressed clay brick and a wire cut clay brick?

The *pressed brick* has sharp arrises with an indentation in the top called a frog, usually with the maker's name stamped on it, and frequently another smaller frog in the bottom or bed surface of the brick.

15

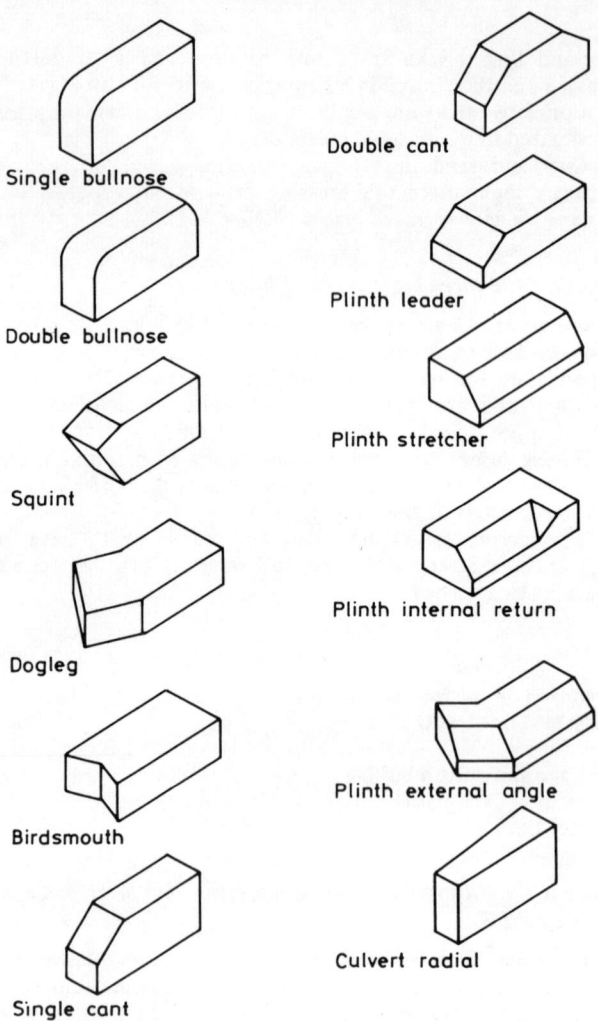

Single bullnose

Double bullnose

Squint

Dogleg

Birdsmouth

Single cant

Double cant

Plinth leader

Plinth stretcher

Plinth internal return

Plinth external angle

Culvert radial

Figure 17. Purpose made special bricks

16

The *wire cut brick* is usually not so regular in appearance: it does not have a frog but wire scrape marks on the top and bed of the brick show where it was cut off by wires from the column of extruded clay during manufacture.

The frogs in pressed bricks range from a deep V-shaped frog in one bed face only, to a very shallow depression with the maker's name on both bed faces. The deep frogged bricks are made to be laid frog down with a pocket of air trapped in the space. This greatly reduces the weight of the wall. However, some loss of sound insulation occurs and an increase in heat insulation. The strength of brickwork built this way is lower than the same bricks built frog up and filled with mortar, but the strength of the wall would still be quite adequate for most normal domestic building.

What different shaped bricks are manufactured for special purposes?

There are a large number of standard specials used in brickwork, but most of them are rarely seen. In addition, special bricks are purposely made for a particular job (*Figure 17*).

What is a perforated brick?

A wire cut brick which has had holes punched through it before firing. This has the effect of reducing the weight, increasing the insulation value and improving the grip of the mortar bedding. Little reduction in strength results from this process and there is some reduction in clay used. More even burning is claimed through the centre of the brick while in the kiln, thus giving uniformity of strength through the brick. The proportion of void spaces varies with the size, shape and number of perforations. When the perforations exceed 50% of the total volume, the brick is known as a *cellular brick*, suitable only for lightweight partitions. Perforated bricks which are twice the width of a normal standard brick have a hand hold web in the centre to allow the bricklayer to pick them up with one hand (*Figure 16*).

How are bricks measured for cutting?

Rough cutting of half-bricks or three quarters is often done without measuring, but all facework cutting should be properly measured. Apart from using a rule or tape measure, a cutting

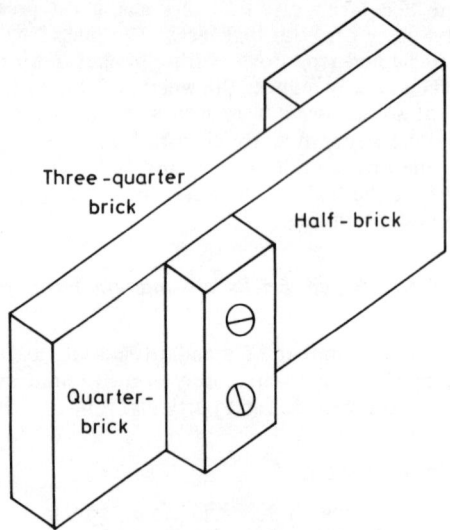

Figure 18. Brick cutting gauge

gauge can be used. This is cut from a piece of wood 75 mm wide and 20 mm thick, with stop, end fixed for measurement of half-, three-quarter and quarter-bricks, as shown in *Figure 18*.

What are the standard cut bricks used in brickwork?

The *half-bat* is 102.5 mm in length (*Figure 19(a)*).
   The *three-quarter bat* is 160 mm in length (*Figure 19(b)*).
   A *quarter-bat* is cut 46 mm length (*Figure 19(c)*).

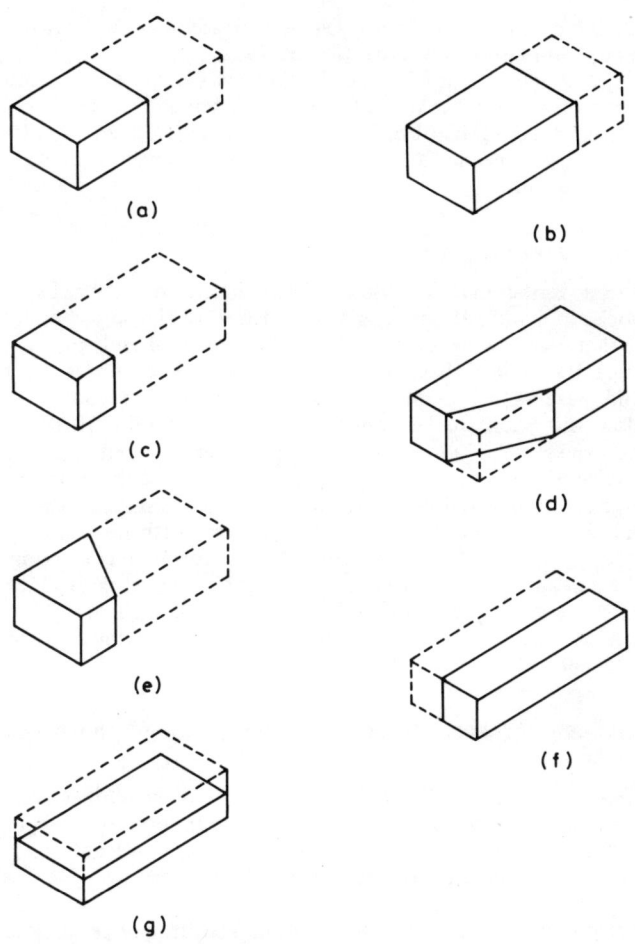

Figure 19. Cut bricks. (a) Half-bat. (b) Three-quarter bat. (c) Quarter-bat. (d) King closer. (e) Mitred bat. (f) Queen closer. (g) Split brick

King closer, cut as *Figure 19(d)*, generally used with *mitred bat* to form a recessed reveal (*Figure 19(e)*).

Queen closer — a brick cut lengthways (*Figure 19(f)*), used next to the quoin header to form quarter bond.

Split brick, a half-brick cut as *Figure 19(g)*, used where a half-course is needed to make up the correct level of the wall, e.g. under sills.

What is a good mortar?

Mortar is the material used to bed and joint the bricks or blocks in a wall. It has to bind well together and spread easily so that the bricklayer can pick up a trowelfull and spread it along the wall with a minimum of effort. The mortar must hold water and remain in a plastic condition long enough to allow the bricks to be adjusted to line and level even where the bricks are of an absorptive type. A good bond must be developed between the mortar and the brick in order to ensure stability and resistance to weather. Resistance to frost and rain and a fairly rapid development of strength are necessary to ensure the long life of the wall and allow continuous working. The mortar should be of a similar strength to the brick or block being used and definitely not stronger than the brick. A building mortar generally consists of natural sand with a hardener and a plasticiser.

What are the main types of mortar in use for brick and blockwork?

Lime mortar is a mix of 1:2 lime and sand which has good working properties if thoroughly mixed, but develops strength very slowly and so is little used in modern construction. It does not set but stiffens slowly as it dries out.

Cement mortar, a mix proportion of 1:3 cement and sand, is required to give a good workable mortar. It may be suitable for high strength engineering brickwork but is too strong for most ordinary work. Leaner mixes of cement and sand become harsh and unworkable.

Cement lime mortars — the addition of lime to a cement mortar will improve the workability considerably and allow a lower cement—sand ratio to be used. A lower strength mortar will thus result, having most of the properties of cement and lime mortars.

Plasticised cement mortar — to improve the workability of weaker cement/sand mixes, air entraining additives are used as an alternative to lime.

Masonry cement mortar consists of a mixture of Portland cement with a fine mineral filler and an air entraining agent. A very fine, easily worked mortar.

How strong a mix should be used with the different brick types?

There are five recognised mortar groups covering all types of mortars. The mixes are grouped according to strength development.

Group	Cement, lime, sand	Masonry, cement, sand	Cement, sand, plasticiser
1	1:0:3		
2	1:½:4	1:3	1:3 or 4
3	1:1:5 or 6	1:4 or 5	1:5 or 6
4	1:2:8 or 9	1:5½ or 6½	1:7 or 8
5	1:3:10 or 12	1:6½ or 7	1:8

Group 1 mortar is only suitable for high strength clay engineering brick.

Group 2 mortar is suitable for high strength Class 1 concrete the sand lime bricks.

Group 3 mortar is suitable for facing brickwork exposed to weather, such as parapet walls and work below dampproof course (d.p.c.). General facing brick and blockwork exposed to weather.

Group 4 for load bearing, internal walls using low strength clay bricks or Class 4 or 5 sand lime or concrete bricks.

21

Group 5, for non-load-bearing internal walls using low strength clay or Class 5 sand, lime and concrete bricks.

How are mortar colours used for decorative architectural effect?

Mortars are naturally various shades of grey due to the cement used. The variation is mainly caused by the different colours of the sands used, although there are slight variations in cement colour. A lime cement mortar always produces a lighter colour than a straight cement sand mix.

Mortars in a variety of colour can be produced by adding colour pigments during mixing. The pigments, either powder or liquid, have to be very accurately measured and thoroughly mixed with the cement and sand to ensure even colouring of all mixes. Colour cements are basic white cement, with pigment added during manufacture. To obtain the best results from colour pigment, very clean light coloured sand or silver sand should be used with white cement. These mortars can be very expensive, and when used for brick and blockwork are used only for pointing after the structural mortar has been raked out.

What materials are used to make concrete?

Concrete is a hard, durable material, having great compressive strength and good wearing and weathering qualities. It can be cast in a great variety of complex shapes, and when reinforced forms one of the main structural materials used in building and civil engineering.

Concrete consists of a coarse aggregate which is generally natural gravel or crushed rock particles greater than 5 mm in size, forming the bulk of the concrete. A fine aggregate, usually natural sand less than 5 mm, is used as a filler to the coarse aggregate. These materials are bound together with a cement paste, which, after setting, hardens to mass into a homogenous material. Water is used to activate the cement powder to form the cement crystals and also to lubricate the mix, making it workable.

How are the materials in a concrete mix measured?

For small quantities of relatively little importance the materials can be measured by bucketful or shovelful, but these methods are unreliable because even with care it is not possible to measure the volume according to its moisture content. This *bulking* of sand, as it is called, means that volume batching is not used where good quality concrete is required. Weight batching is the method most widely used to produce high quality concrete.

Each material is weighed in accordance with the mix proportions required, the water being measured with equal accuracy.

When the dry materials are mixed to make concrete there is a considerable loss of volume, usually about 40%. This is because the finer particles fill the void spaces in the coarse particles.

How is concrete placed and cured to achieve maximum strength?

Concrete sets within ten hours but hardens and gains strength continuously, however, so that almost maximum strength is reached at 28 days. To attain this strength it is essential to remove all the air from the concrete during placing.

Methods of compaction vary. Hand compaction is suitable for small jobs but requires a wetter, more workable concrete than mechanical compaction. Mechanical vibrators can be clamped on to formwork constructed to contain the concrete in the desired shape. They shake the air out of the concrete. Poker vibrators are used with the same effect, but are inserted into the concrete at 0.5 m spacing.

Beam vibrators are used to tamp concrete which is laid in flat areas, such as floors. In all cases, if concrete is to harden properly then it must not be allowed to dry out. Keeping the concrete damp is known as *curing*. In warm weather the concrete must be cured for 7–14 days. Methods of curing include covering with damp cloths which are kept damp by spraying with water as necessary. Similarly, sand or straw covering can be used. Polythene sheeting or similar waterproof materials will also prevent evaporation of water from concrete and do not need further attention. Waterproof resin membranes painted on are also used, mainly on roads.

3

WALLING

What is meant by the term bonding?

Bonding is the arrangement of bricks or blocks to a definite pattern in the wall.

There are three reasons for bonding. One is to give the wall structural strength by ensuring that no vertical straight joints

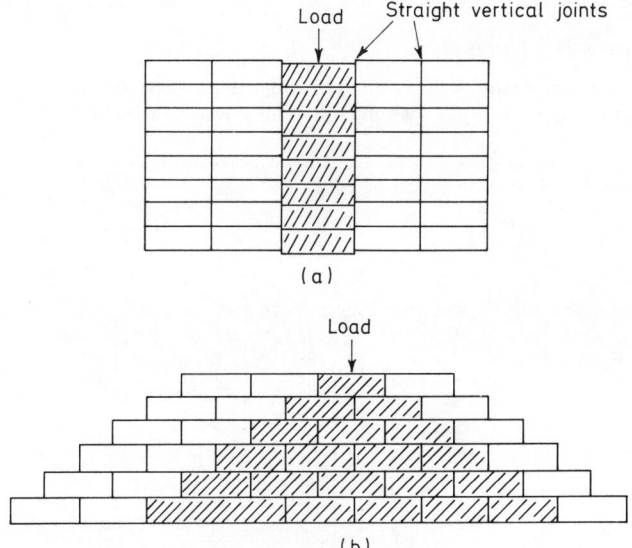

Figure 20. Bonded and unbonded walls. (a) Uneven load distribution in an unbonded wall. (b) Distribution of load through a bonded wall

appear. An unbonded wall having continuous vertical joints would have vertical planes of weakness and under load would tend to crack much more easily than a bonded one (*Figure 20*).

An aesthetically pleasing and decorative appearance can be given to a wall by using a bonding arrangement which will form a pattern feature in the wall. Patterns may range from the very simple, incorporating headers into a mainly stretcher face wall, to involved arrangements using different colour bricks and intricate patterns.

Economy may play an important part in selecting a facing bond. Bonding patterns which use stretcher faces only are the most economical because there is less pointing and bricks are easier and quicker to lay as stretchers. By introducing headers into the wall, extra mortar joints have to be pointed on the face and more facing bricks are required.

How many bonds are there?

There are many bonds, but basically all are variations on the basic half-lap bond, where all bricks lap half a brick over

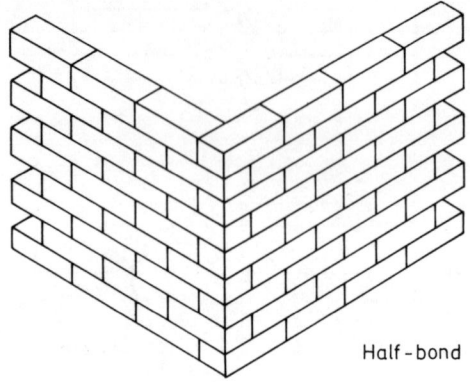

Half-bond

Figure 21. Stretcher bond

26

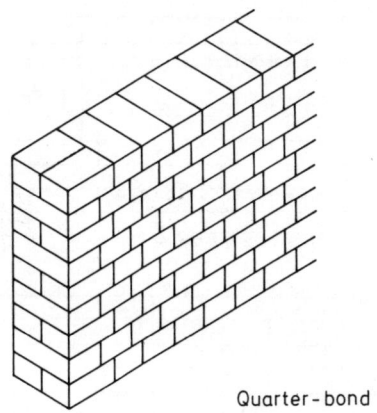

Quarter-bond

Figure 22. Header bond

the other bricks in the course below (*Figure 21*). There is also quarter-lap bond, where the bricks lap over a quarter of a brick length (*Figure 22*).

Which bond is the most commonly used?

Because walls half-brick thick are the most common types used in building, stretcher bond is the most common. This is simply a half-lap bond where all bricks present their stretcher faces to the face of the wall (*Figure 21*).

Which is the strongest bond?

It is generally accepted that for walls of one brick thickness and over, English bond is strongest, as it forms no straight joints of any kind within the thickness of the wall (*Figure 23*). This bond is used on all work where strength and load bearing properties are most important. The quarter-bond is obtained by inserting a queen closer next to the first header in each header course.

27

Figure 23. English bond

Figure 24. Flemish bond

Which is the main decorative bond used in brickwork?

Flemish bond (*Figure 24*), a pattern consisting of alternate headers and stretchers in each course. When using in a half-brick wall or the outer skin of a cavity wall, half-bats are used for the headers. To increase the decorative effect, the headers are sometimes a contrasting colour or texture. Further effect can be obtained by projecting or setting back some of the headers into the wall.

What are garden wall bonds?

Modifications of standard English or Flemish bonds, introducing more stretchers into the wall and reducing the number of headers but retaining the basic bond patterns. Originally they were used for boundary walls of one brick thickness when hand moulded and burnt bricks were in general use. These bricks were very irregular in shape and size because of the method of manufacture, and it was difficult to select bricks of uniform length with two good header faces to use as headers in one-brick facing walls. The garden wall bonds reduced the number of bricks required for headers and were more economical.

English garden wall bond is still used for reasons of economy and consists generally of three or five courses of stretchers to one course of headers (*Figure 25*). The stretchers are easier to lay and there are fewer crossjoints to be pointed.

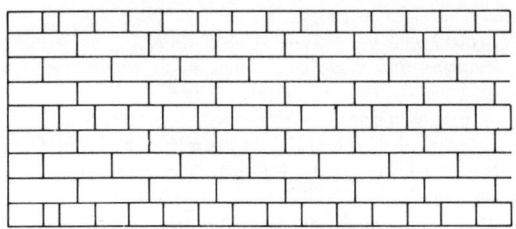

Figure 25. English garden wall bond

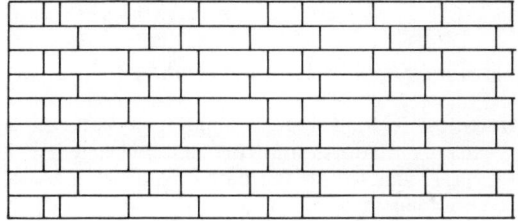

Figure 26. Flemish garden wall bond

Flemish garden wall bond, consisting of a pattern of three, four or five stretchers to one header (*Figure 26*), forms an economical pattern effect on the outer skin of a cavity wall. The snap headers can be of a contrasting colour or either inset or projecting.

What terms are used to describe brick walling?

1. *Bed joints* are the horizontal mortar joints in a wall.

2. A *course* (of brickwork) is one row of bricks plus the bed joint.

3. *Lap of bond* is the horizontal distance between the vertical joints of two successive courses.

4. *Perpends* or *crossjoints* are the vertical joints in a wall.

5. *Quoin*, the external angles of walls.

6. *Quoin header*, the first header at the end or corner of the wall.

7. *Quoin stretcher*, the first stretcher.

8. *Racking back*, setting back each successive course of brickwork to be built up at a later stage.

9. *Return*, a corner.

10. *Stopped end*, the finished plumb end of a wall.

11. *Straight joint*, two perpends occurring vertically in the face of the wall.

12. *Toothing*, each alternate course overhangs the one below by a distance equal to the lap. It is a method of leaving the end of a wall which is to be extended at a later date.

What is meant by racking back and toothing?

When a corner is built before building a wall, the work is stepped back each course until the required height is reached (*Figure 27(a)*). This is also the best way to leave a wall which is to be extended at a future date. However, where it is not possible to run a wall out sufficiently to rack it back, a method known as *toothing* is used, where every other course is stepped over (*Figure 27(b)*). It is more difficult to ensure strong solid bonding in with this method when continuing the wall. To make certain the racking back is kept in line with the wall, the level or straight edge is held at an angle down the wall as shown in *Figure 27(a)* and the bricks straightened to it. The end of a toothing is plumbed in the normal way.

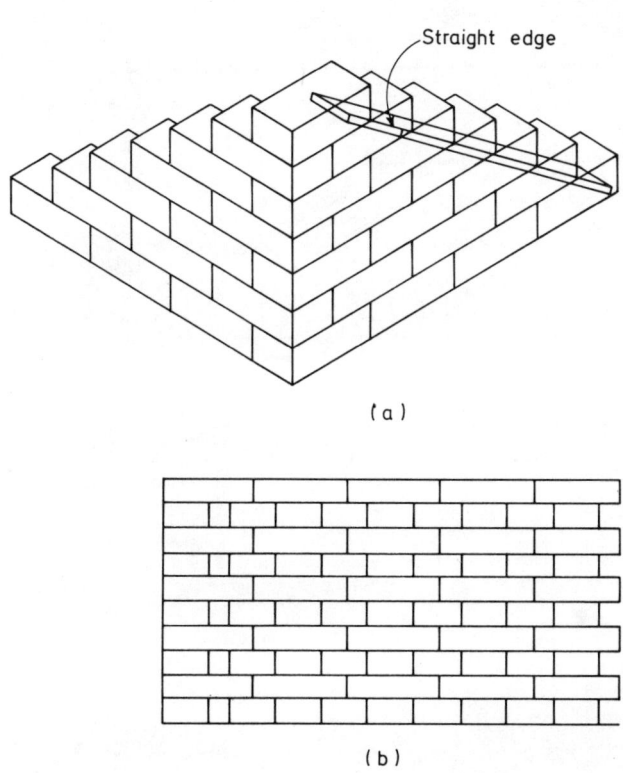

Straight edge

(a)

(b)

Figure 27. (a) Racking back. (b) Toothing

What is the method used for building walls straight, level and plumb?

The corners are built first. Care must be taken to ensure that they are built plumb and to gauge also that the courses are level. The work in between the corners is kept straight by using a line fixed to the corners at each end. This line is

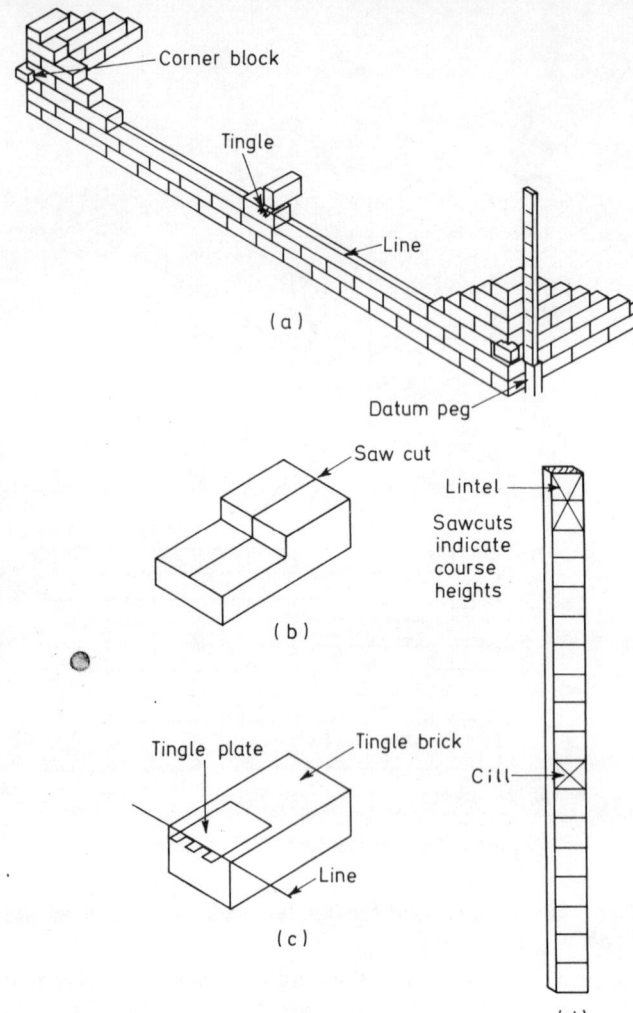

Figure 28. (a) Use of a line, corner block, tingle and gauge lath. (b) Corner block. (c) Tingle. (d) Gauge lath

32

raised a course at a time and the bricklayer keeps the top arris of each brick to it, ensuring a straight and level course. The line is fixed at each end either by line pins or corner blocks. The pins are pushed into the appropriate crossjoint or taken round the corner as shown in *Figure 28(a)*. The corner block is a wood block 75 × 50 × 50 mm, cut as shown in *Figure 28(b)*. The line is tightened through the saw cut and the block can be raised a course at a time, as with the line pins; the tightness of the line holds the block in position. The advantage of blocks is that there are no pinholes left in the wall to be made good.

If the wall is a long one, the line is liable to sag in the middle. In windy weather it may also be moved by the wind, making it difficult to lay bricks. To prevent this, one or more *tingles* are used, depending on the length of the wall. A tingle is a metal plate which holds the line in position at an intermediate place along the wall (*Figure 28(c)*). To make certain that the tingle is in the correct position the tingle brick can be plumbed and gauged or the line itself can be sighted through from one end to the other.

What is the gauge of brickwork?

Gauge is the vertical measurement of each course in a wall. The correct height of brickwork can be maintained by using a wooden gauge lath (*Figure 28(d)*). This lath is marked with saw cuts at intervals of one course. When used as shown in *Figure 28(a)* the bricklayer can check the level of each course and also be sure that the corners are built to the required height. Additional information can be shown on the gauge lath such as sill heights, lintel heights and other features of the building.

How is a small rectangular building set out in the correct position on the site?

The position of a building is set out on the ground by using wooden pegs and lines. First, the frontage or building line is

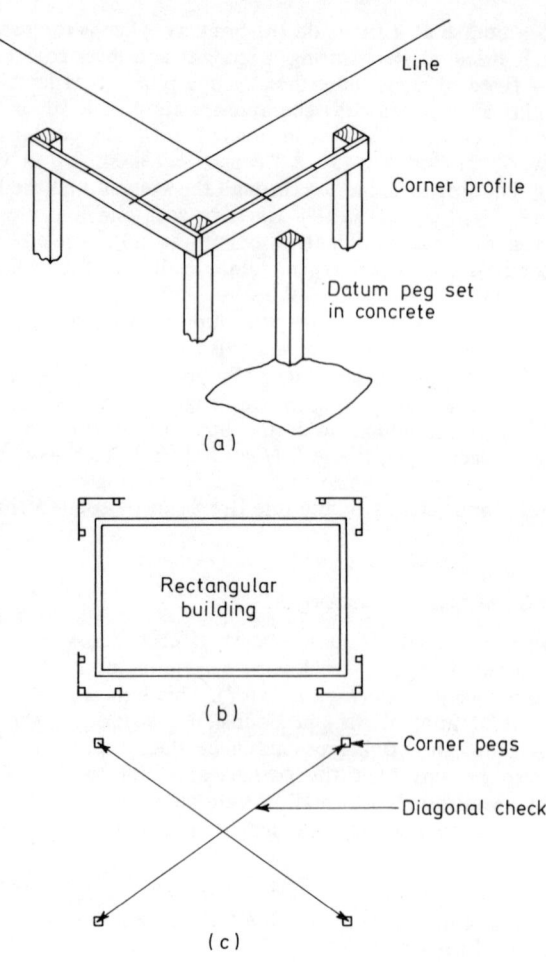

Figure 29. (a) Fixing a corner profile. (b) Four corner profiles. (c) Method of checking accuracy of corner profiles

34

set out. This line is fixed by the local authority and is usually a measurement from the crown of the adjacent road. No part of the building may be in front of this line. A cotton setting-out line is stretched between two pegs across the frontage of the site to show the position of this line. The left-hand front corner is then measured and a wooden peg driven into the ground along the frontage line; for accuracy a nail is driven into the top of the peg at the exact point (*Figure 29(a)*). This may be measured from an existing feature shown on the plan. Having fixed the position of the front left-hand corner it is easy to measure along the frontage line and fix the position of the front right-hand corner with a wooden peg and nail as before. Next, the rear left-hand peg is measured and fixed square from the first peg, and lastly the rear right-hand peg is measured and driven.

There are now four pegs, each with a nail in the top, marking the exact position of the four corners of the building (*Figure 29(b)*). These can be checked for accuracy by measuring the diagonals, which should be equal (*Figure 29(c)*).

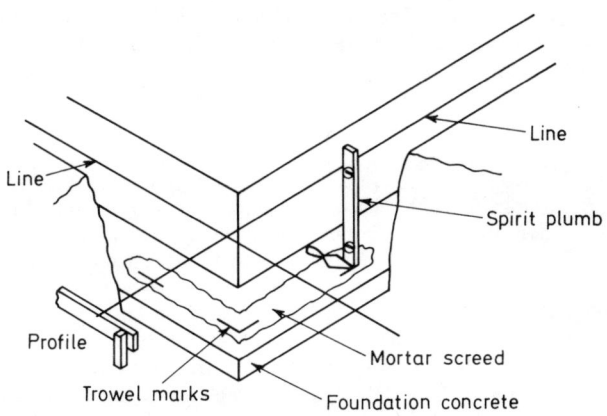

Figure 30. Transferring setting-out lines on to concrete foundation. Side of trench is cut away to show foundation concrete

How is the position of a building indicated to the builders?

As soon as excavation for the building foundation commences, the corner pegs will be dug out so corner profiles are set up at each corner where they will not be disturbed by excavation. These profiles are wood boards nailed to pegs and with saw cuts or nails driven in to indicate the position of the building and to which lines can be fixed (*Figure 29(a)*). The position of intermediate walls can also be fixed by setting up profiles. Lines are fixed to these profiles and the foundation trenches can be dug out; also the brick foundation walls set out from them. A thin cement sand screed is trowelled on to the foundation concrete at the corners of the building and the ranging lines plumbed down and marked with the trowel point as shown in *Figure 30*.

How are the levels for the building determined?

A level or datum peg is driven at each corner of the building. The top of the peg is set at a predetermined level, i.e. d.p.c. level or floor level. The depth of foundation trench can then be measured down from this. Datum pegs are often painted so that they can be identified easily by their colour, and are set in concrete to prevent disturbance (*Figure 29(a)*).

What are the foundations of a building or wall?

The foundation is the base of the wall, generally built down into the ground. The foundation is taken down to a depth where an undisturbed subsoil will be unaffected by changes in the weather, by drying out, saturation or frost. The load which the wall will place on the foundation and the bearing capacity of the subsoil are the two factors to be taken into account when determining the width of the foundation.

The concrete foundation gives a hard level surface for the brick or blockwork to be built on and spreads the load of the wall over a greater area of subsoil. On sloping sites, this concrete is stepped down in a series of steps which correspond with

36

brick or block courses, thus presenting the bricklayer with a level surface to work from.

The Building Regulations lay down the minimum standards required by law for the foundations of domestic buildings. The thickness of the concrete must not be less than 150 mm or not less than its projection from the base of the wall if this is greater than 150 mm (*Figure 31(a)*). The width of the

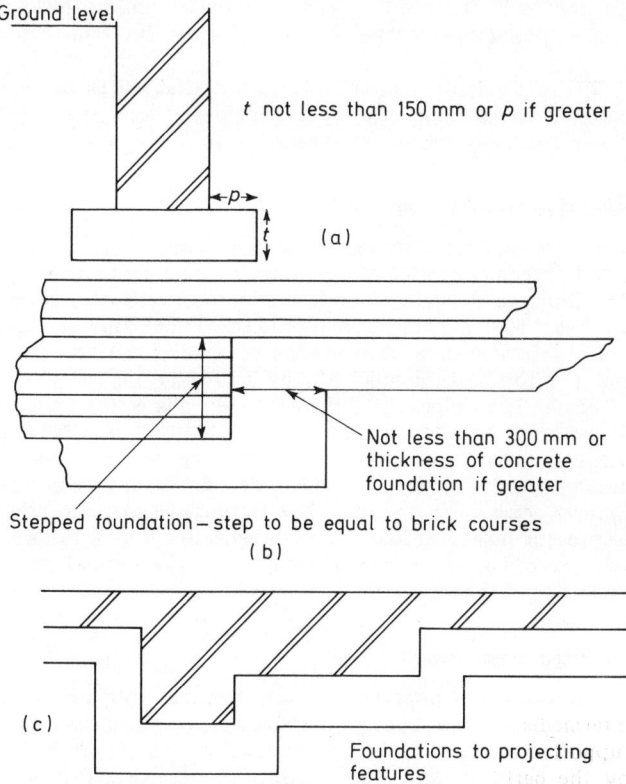

Ground level

t not less than 150 mm or *p* if greater

←*p*→

↕*t*

(a)

Not less than 300 mm or thickness of concrete foundation if greater

Stepped foundation – step to be equal to brick courses

(b)

(c)

Foundations to projecting features

Figure 31. Minimum dimensions for concrete foundations

foundation is specified in a table which identifies the types of subsoil and its condition and gives minimum widths for a range of wall loadings. The concrete must be a mix proportion of 50 kg cement, 0.1 m^3 of fine aggregate, 0.2 m^3 of coarse aggregate.

Where a foundation is stepped on a sloping site at each change of level, the higher level foundation must extend over and join with the lower level by not less than the thickness of the foundation and in no case less than 300 mm (*Figure 31(b)*).

Where a pier or projecting feature is attached to the wall, the foundations must project on all sides at least as much as they project beyond the wall (*Figure 31(c)*).

What is the oversite concrete?

A slab of concrete laid on to the compacted hardcore in the area between the outer walls and under a hollow timber floor. The Building Regulations state that the concrete shall be not less than 100 mm thick, composed of cement, fine aggregate and coarse aggregate. A trowelled or spade finish is suitable and the top surface must not be below the highest ground or paving level adjacent to the external walls of the building. There must be a clear space not less than 75 mm between the top surface of the concrete and any wall plate, and a minimum of 112 mm to the underside of any suspended timbers, such as floor joists. The purpose of this concrete is to prevent plant and fungal growth beneath a hollow floor and to prevent moisture vapour rising from the ground to the underside of the timber floor.

What is a sleeper wall?

A wall erected at intervals between the main walls to provide intermediate support for ground floor joists. The joists may be supported at each end by the inner skin of the cavity wall or by the partition wall. The intermediate sleeper walls reduce the joist span and allow smaller section timbers to be used, thus reducing cost considerably. Sleeper walls are usually built up on the oversite concrete and are a honeycombe construction designed to allow free circulation of air under the floor.

What is the fender wall?

A wall built to contain the supporting hardcore and carry the concrete hearth to a fireplace in a timber ground floor. The wall may also be combined with a sleeper wall supporting the joists of the floor. A d.p.c. must be built into the wall at the same level as the other sleeper walls. A fender wall is not necessary to a fireplace in a building with a solid concrete ground floor.

Why are ventilators built into the external walls below a hollow timber floor?

Timber is attacked by a fungus growth known as *dry rot* in conditions where there is some moisture and still air. These conditions can exist under a timber floor and the fungus can destroy it in a very short time. It can also penetrate other timber away from the floor, while deriving its moisture below floor level. A circulation of fresh air will prevent this growth and is provided by placing ventilation air bricks in the outer walls and by honeycombing any intermediate support walls. The ventilators in a cavity wall should incorporate a duct to ensure that the air circulates under the floor and does not escape into the cavity.

What is a wall plate?

A timber plate bedded on to a wall which is going to support joists. Usually, wall plates are bedded with mortar on to a d.p.c. laid on the sleeper walls below a timber ground floor. The wall plate should be treated with a preservative and the joists nailed to it. A wall plate is also used to eaves level to support the roofing timbers.

What is a dampproof course?

A layer of impervious material incorporated in a building to resist the passage of moisture. Materials used for this purpose are of two types.

1. *Rigid materials.* Two courses of slates or tiles laid in cement mortar 1:3 bonded to prevent straight joints. This d.p.c. can be used both vertically and horizontally but is mainly used with a brick on edge course to form a coping on a free-standing wall. Class A engineering bricks in cement and sand mortar 1:3, a minimum of two courses, are used as a horizontal d.p.c. where very heavy loadings are expected.

2. *Flexible d.p.c.s.* Sheet metals of the non-ferrous type, lead, copper, zinc and aluminium, are used mainly in flashings and as d.p.c. in walls where heavy loadings are expected. Bitumenous sheet materials are the most commonly used as they are the cheapest. The bitumen is formed into a sheet material by using a hessian or asbestos fibre as a base; some types have a lead foil core incorporated in the centre of the sheet. Plastics sheet materials, polythene or p.v.c., are thinner sheet than bitumen, very tough and durable.

The bitumen and plastics d.p.c.s are produced in rolls of varying width to suit builders' needs. They are very easy to use and cut, though the bitumenous types have to be warmed in very cold weather as they get brittle and crack when being unrolled. The rigid types are liable to crack if there is the slightest movement in the wall and let moisture through.

Why do the Building Regulations state that a d.p.c. must be built into every wall, forming part of a building at a minimum height of 150 mm above ground level?

The purpose of this d.p.c. is to prevent ground damp rising up the wall and causing damp areas to appear on the ground floor walls of a building. Most bricks, blocks and stones are porous and will absorb moisture by capillary action through the pore spaces. Foundation walling is always in contact with damp ground and moisture is drawn up the wall by capillarity in the pore structure of the walling material. The horizontal d.p.c. is designed to provide an impervious barrier and prevent moisture rising into the building. 150 mm above ground level is specified as a minimum to take account of rain splashes from paths and possible changes in the ground level.

What is a dampproof membrane?

D.P.M. is a damp resistant material incorporated in a solid concrete floor. It is usually sandwiched in the floor between the structural concrete floor and the screed topping. Sheet materials such as p.v.c., bitumen, polythene or building paper, which is a layer of bitumen between two sheets of paper, are commonly used for this purpose, but liquid bitumen and rubber compounds can be used applied by brush or squeegee. The sheet types can be laid on to hardcore and the concrete floor laid on top, but considerable care is needed to make sure the sheet is not punctured by sharp edges in the hardcore. The d.p.m. must be joined to the horizontal d.p.c. in the walls at the edges of the concrete floor.

What is a cavity wall?

Cavity walls were introduced at the beginning of this century to overcome the problem of dampness appearing on the inside of the external walls of buildings. As most brickwork and masonry materials are porous, moisture was able to find its way through solid walls and cause damage and discolouration to the plaster and decoration on the inner surface of the wall. To prevent or reduce this, walls had to be very thick and therefore expensive. The cavity wall consists of two separate skins with an air space of at least 50 mm between them. The thickness of these skins may not be less than 100 mm but they can be of different materials, e.g. facing brick outer skin and lightweight blockwork inner skin.

The two skins are tied together by means of metal ties, set out in a regular pattern and staggered. The ties are twisted in the middle to prevent the passage of water across and must be rustproof. In order to be successful, care must be taken to ensure that the cavity is kept clean and free from mortar droppings, which may build up on the wall ties and provide a bridge for moisture to cross from the outer to the inner skin. At any point where the cavity is closed, a d.p.c. is incorporated to prevent the passage of moisture. Examples of this are at window and door openings.

What types of wall ties are in general use for cavity walls?

Wall ties are made from galvanised steel wire or galvanised steel strip. Although several patterns are made, they must conform to BS 1243. The steel strip types have fish tail ends to ensure a strong grip in the mortar bed, and a twist in the centre which will make sure that any moisture penetrating the outer skin will not be able to travel across the cavity. The wire ties have twist or bend in the centre to perform the same function. Non-ferrous metals can also be used for wall ties.

What are the advantages of a cavity wall over a solid wall?

The cavity wall will prevent moisture penetrating a building. To ensure this a solid wall would have to be very much thicker, and/or built with an impervious brick or stone, e.g. Class A engineering brick or granite. In either case, the solid wall would be more expensive, thicker and heavier than the cavity wall. Two very dissimilar materials can be used in the skins of a cavity wall which would not be possible in a solid wall. Generally, a cavity wall can give better heat insulation than the same thickness solid wall. Decorative work can be carried out on one skin of the cavity wall, while common work is carried out on the other. The outer skin is the weather resistant wall, the inner skin can be varied in thickness and material to provide for load bearing and insulation as necessary.

How is a cavity wall built?

When working from the outside of the wall, the inner skin is built up first to six courses of bricks or two courses of blocks,

450 mm. The line is then transferred to the outer skin and that is built up to the same height. The wall ties are laid on the wall at 900 mm centres and the inner skin taken up another 450 mm. This process is repeated until the desired height is reached.

To prevent the build-up of mortar on wall ties, a cavity batten is used. This is a timber batten, slightly less than the width of the cavity, laid on the wall ties at each level. As each wall tie height is reached, it is pulled up, thus bringing any dropped mortar with it. The batten is then cleaned off and replaced on the next layer of wall ties. Other methods of keeping the cavity clean are to use a long batten with a piece of cloth fastened to the end, and at the end of each day's work, to clean off the wall ties with it. This method is not suitable for buildings more than two-storey. Coring holes can be used above windows and door openings. These are cleaning out holes left in the outer skin of the cavity wall, above the lintels, thus giving access to the d.p.c. over the lintel and an opportunity to clean it off if necessary.

What is the purpose of concrete filling to the cavity at the base of a wall?

The fine concrete 1:2:4 mix using 10 mm aggregate is tamped into the cavity to increase the strength of the wall below floor level. The inner skin of a cavity wall at this point is subject to considerable side pressure when the hardcore and the concrete floor or oversite is compacted. The concrete filling forms a solid wall and stabilises the foundations, enabling the hollow wall to resist these pressures and any side pressure exerted by the external ground. The top of the concrete must be not less than 150 mm below the horizontal d.p.c.

How is the cavity kept clean and free from mortar droppings during construction?

One method is to use the cavity battens or sweeps. These are wood battens placed on the first level of wall ties which catch the mortar droppings as the two skins of the wall are built up to the next wall tie height. Before placing the next row of wall ties, the batten is lifted out by means of the wire or cord handles provided, cleaned off and replaced on the wall ties and the process repeated. At the end of a day's work or when a scaffold height is reached, the wall ties can be cleaned off with a long timber lath should this be necessary.

At the base of the wall, below d.p.c. level, holes may be left at intervals to allow for cleaning out the base of the cavity. Similar cleaning-out holes can be left above door and window openings or any other place where the cavity is bridged to allow for cleaning any surplus mortar droppings off the d.p.c. These holes, usually one brick long and two courses high, have to be made good when the wall has reached its full height.

How are cavity walls finished at eaves level?

The cavity is closed at eaves level, either by laying a course of headers across the cavity or by bedding a course of concrete slabs. The wall plate for the roof is then bedded on to this. No damp course is required at this level as the overhang of the roof and guttering prevents the top of the wall from getting wet.

What are gables?

Gable walls form the ends of a pitched roof. They form a tri-angular shape and involve raking cutting on both sides to the pitch of the roof.

Two methods are usually adopted to construct these walls to the correct angle. A temporary pattern spar may be set up by the carpenter which the bricklayer may use to cut the wall to the angle. A more accurate method of facework cutting is to have a timber framework set up which will support a line allowing the bricklayer to mark and cut the bricks or blocks to form the gable.

Gable walls are not very stable — standing on their own they are very easily blown down during construction. It is

44

essential that there is close cooperation between the carpenter and the bricklayer to ensure that the gable walls and roof timbers are constructed together so that a stable structure can be formed as soon as possible and the roof covering fixed. In some types of construction, the roof timbers can be fixed first and the gable walls built up afterwards.

At the junction of the gable and the main walling at eaves level, a corbel is usually built in, known as a *knee*. This corbel may be in the form of a simple precast concrete or brick corbel, or can be made into a decorative feature by incorporating tiles, dressed stones or coloured bricks.

How is the cavity wall finished at the verge of the gable?

The tiling or slating to the roof may be finished flush with the brickwork or with a small overhang. Roofing battens are nailed to a rafter which is fixed into the cavity of the gable wall. The slates or tiles are nailed to this and bedded in cement mortar on to the outer skin of brickwork. Projecting feature courses of brickwork are sometimes used to support the roof covering at the verge. Where the roof projects beyond the gable wall, corbel timbers are built in to carry the end rafters and roofing.

What methods are used to fix timber frames into brick jambs?

During construction of the jambs of an opening, wood pads can be built into the bed joints of the brickwork. These pads are pieces of preservative-treated timber 100 mm square by 10 mm thick. The first pad is built in 150 mm above the bottom of the frame and the last one 150 mm from the top. Intermediate pads are built in at approximately 600 mm intervals. The frame is then firmly fixed by nailing through into the timber pads.

An alternative method is to build fixing bricks into similar positions to the timber pads. Fixing bricks are composed of clinker or similar lightweight concrete which will take nails. When the frames are to be built into position, fixing cramps can be screwed to the back of the frame and built into the brickwork. The cramps are galvanised steel with fishtail ends and are spaced as wood pads. An alternative method is to use two 100 mm nails in place of a metal cramp. Metal windows and door frames usually have fixing cramps specially made for the frame, with slotted holes which allow adjustment to suit the bed joints of the brickwork. Hardwood frames which are fixed at a later stage of construction can be screwed to the wood pads or fitted with cramps which are then built into holes left in the brickwork during construction of the walls. A plastics sub-frame, complete with fixing cramps, can be screwed to the timber frame and built into the cavity of the jamb. This acts as a cavity closer, vertical d.p.c. and frame fixing.

4

BLOCKS AND BLOCKWORK

What types of concrete building blocks are in general use?

Blocks are walling units larger than that specified for bricks. The heights must not exceed the length or six times the thickness. Concrete blocks are classified as Types A, B, and C by BS 2028.

Type A are dense aggregate blocks intended for load bearing walls.

Type B are lightweight aggregate concrete blocks for load bearing.

Type C are lightweight aggregate concrete blocks for non-load-bearing walls.

Types A and B can be used in the following positions:

1. External walls protected by cement rendering or some similar finish.
2. The inner leaf of cavity walls.
3. Backing up to brickwork, masonry or cladding panels.
4. Internal load bearing partitions.
5. In-filling to steel framed or concrete framed buildings.

Type C blocks are suitable only for

1. Internal non-load-bearing walls and partitions.
2. Internal non-load-bearing infil panels to framed buildings.

The blocks may have plain ends or be tongued and grooved or double-grooved to improve the key. Dimensions should be such that the blocks will bond with the standard brick courses. Blocks are usually equal to three courses of brickwork in height and two bricks in length.

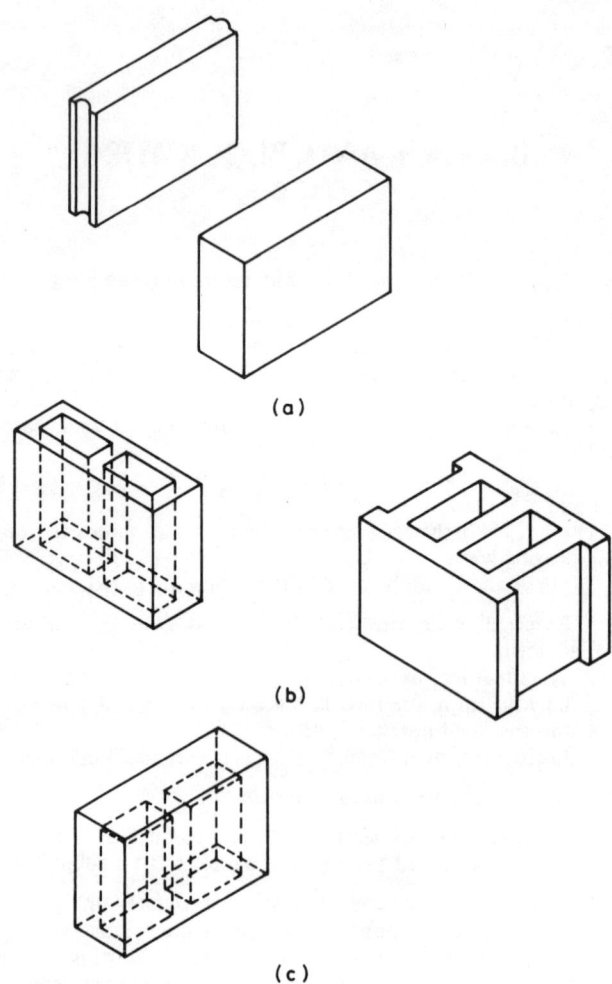

Figure 32. (a) Solid blocks. (b) Hollow blocks. (c) Cellular blocks

48

What is the difference between solid, hollow, and cellular blocks?

Solid blocks are plain rectangular units, with or without tongues and grooves (*Figure 32(a)*).

Hollow blocks have two, three or four vertical cavities which go right through the block (*Figure 32(b)*). They are shell bedded with mortar, which means the mortar is laid along the two outer edges of the block and the crossjoint is applied to the outer edges of the end of the block. This method means that there is a continuous gap in the mortar joint, both horizontally and vertically, improving resistance to rain penetration. Hollow blockwork with two vertical cavities has a continuous vertical cavity which can be filled with concrete and reinforced vertically to produce a solid reinforced wall.

Cellular blocks have two, three or four vertical cavities which do not penetrate right through the block. This type is laid with the openings downward, which means that a full bed is spread on the block and the cavities do not run continuously up the wall (*Figure 32(c)*).

Are brickwork mortars suitable for blockwork?

The same general rule applies to both brick and blockwork. The mortar should be of a similar strength to the blocks being laid, and never stronger. The dense load bearing blocks require a mortar of 1 cement:4 sand; this will also resist severe weather conditions. For normal external work subject to severe exposure a 1:6 mortar is suitable. A 1:8 mortar should be used for internal walls of medium-to-low strength blocks. Generally, if there is doubt about the mortar mix to be used with any particular block the manufacturer will specify suitable mixes.

Why should concrete blocks be kept dry before use?

All concrete materials are subject to drying shrinkage, and allowance has to be made for this when designing concrete blockwork. The drying shrinkage on blockwork can be kept to a minimum level if the blocks are kept dry before laying and allowances made in construction for the known shrinkage.

If the blocks are allowed to get wet before laying, the shrinkage on drying will be greater than that allowed for by the architect, resulting in vertical cracks in the weakest part of the wall. If the wall has windows and doorways then cracking will occur above one of these openings. To avoid these cracks the walls should be divided into rectangular panels by introducing movement joints into them. Storey height door frames and window panels can be used effectively to break up long lengths of partition walling. Other methods used generally mean building in a movement joint at a suitable position. This joint must run the full height of the wall and may be a dry joint or caulked with an elastic non-hardening compound. The joints may be cloaked with a metal or wood cover strip. Joints in freestanding walls can be masked by incorporating the joint within attached piers which allow the wall to move independently of the piers.

What decorative blocks are available?

Facing blockwork is used extensively as an alternative to, or complementary to, facing brickwork and natural stone masonry.

Special concrete facings are usually dense blocks, or lightweight with a dense facing to resist the effects of the weather. They may have a heavily profiled face to a geometric pattern or be made to resemble natural rock face; they may be split

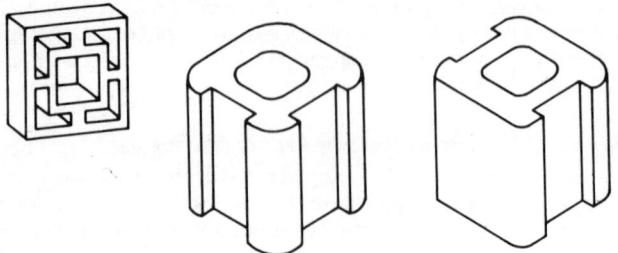

Figure 33. Typical screen and pier blocks

faced or finished with exposed aggregate to give interesting colour and texture. The blocks may be all the same size or in a number of sizes which, when mixed together, give a random effect. Reconstructed stone blocks are made which have a facing of natural stone dust and coloured cement, so that they closely resemble the natural stone, though they have a tendency to lose colour during weathering. Pierced screen blocks of various patterns and colours are available. They are non-load-bearing and are used for panels or in conjunction with a special grooved pier block, plinth or coping. (See *Figure 33.*)

What are aerated concrete blocks?

Aerated concrete is made from cement and a very fine aggregate such as pulverised fuel ash (p.f.a.), mixed together with a chemical aerator. It is cast in large slabs and quickly forms a sponge-like microcellular structure. This slab is sawn into block sizes which are then subjected to high pressure steam in an autoclave for up to 18 h. Blocks of this type do not have tongue and groove ends and are not manufactured as hollow units. They are easily cut and sawn, nails and screws can be fixed direct and they provide good thermal insulation.

Lightweight load and non-load-bearing, they need the protection of rendering if used externally and are unsuitable for work below d.p.c. level.

What aggregates are suitable for concrete blocks?

Type A blocks can be made from the following aggregates:

1. Natural gravels, sands, and crushed rock conforming to the requirements of BS 882.
2. Clean, crushed, clay brick or tile.
3. Granulated slag which contains not more than 50% lime and has a density not exceeding 1900 kg/m³.
4. High density air cooled blast furnace slag.

Type B blocks require lightweight aggregates with a low density. Suitable materials are:

51

1. Foamed slag.

2. Clinker from furnaces which are efficient enough to burn all the combustible material in the fuel and have an inert material.

3. Low density granulated slag.

4. Expanded clay, shale or slate, which has been subjected to rapid intense heating, subsequently expanded into a cellular structure.

5. Natural lightweight aggregates of volcanic origin such as pumice, which is clean and free from impurities.

6. Sintered, pulverised fuel ash which is the ash residue from power stations, using pulverised fuel. The material is in nodule or pellet form or can be used in its powder form as a fine grade aggregate. Also known as fly ash and p.f.a.

7. Chemically treated sawdust and wood shavings combined with some natural sand.

8. Manufactured proprietary lightweight aggregates, mainly based on p.f.a. and forming hollow nodules.

Type C blocks are manufactured from aggregates similar to those recommended for Type B.

How are screen block walls constructed?

Pierced screen blocks are non-load-bearing and have to be supported by piers or some similar structure to prevent them being overturned. They are normally manufactured to form 300 mm square units, then built into a wall with 10 mm mortar joints of thickness 75 or 90 mm. The blocks are built in panels between supporting piers of brick or purpose-made concrete blocks. Steel reinforcing is incorporated to stiffen the panels and increase the wind resistance, also to tie the blockwork panels to the piers. Vertical reinforcement rods bonded into the concrete foundation are built into the piers, and where hollow purpose-made concrete blocks are used for the piers they are filled with concrete. Pierced screen blocks may be used internally between the columns of a framed building to divide space but allow light and air through. Externally, they are used to screen gardens and patios and to provide a wall which will allow air and sun to penetrate but still act as a boundary. The screen block can be obtained as a solid unit with the screen pattern shown on the face to match pierced blocks but give a solid appearance. (See *Figure 33*.)

What are hollow clay blocks?

Perforated burnt clay blocks, laid with the perforations running through the blocks horizontally. The blocks are equal to three courses of brickwork and are 300 mm long. The size

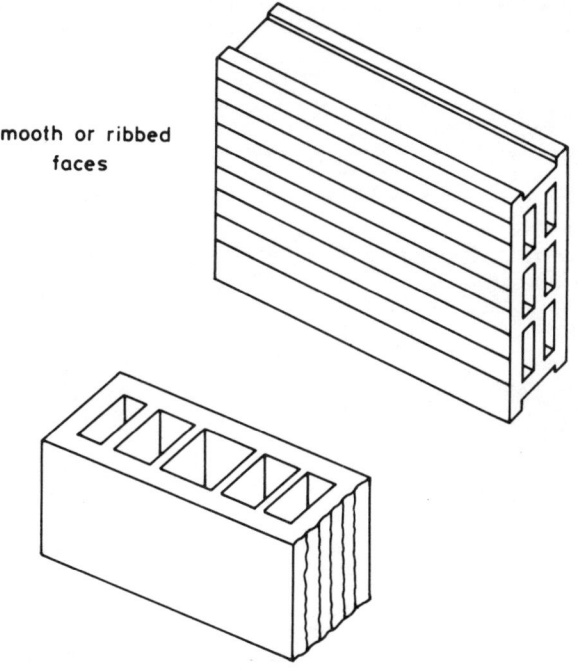

Smooth or ribbed faces

Figure 34. Hollow clay blocks

of the perforations and their number varies with the width of the block and its load bearing capacity. May not be used externally, but provide good sound and heat insulation. They are difficult to cut and chase but special grooved blocks and half-blocks are provided. Can be fairfaced for direct application of decoration or keyed for plaster. They have high fire resistance and do not suffer from the moisture movement and drying shrinkage associated with concrete blocks. (See *Figure 34*.)

5

FREESTANDING WALLS

What is a freestanding wall?

Any wall which is not part of a building. It carries no loads and does not have any restraint upon it. Typical examples are boundary walls, garden walls, decorative screen walls, partition walls not fixed to the structure and parapet walls.

If the wall is external then it will be exposed to the weather on both faces and the top. It will therefore have to be built up from materials which will stand the effects of weather and will also present a good face appearance. Other essential features of this type of wall would be a d.p.c. built in above ground level and a coping to protect the top of the wall. A freestanding wall within a building would not need these features but all would need to be stable enough to resist the overturning forces of wind pressure and people leaning against them. Stability is based upon the slenderness ratio, i.e. the ratio between the effective height and thickness of the wall. Exposed freestanding walls must not be rendered on both sides.

What is a parapet wall?

A wall situated at roof level on a building which rises above the eaves or verge of the roof. Parapet walls are generally built in conjunction with flat roofs (*Figure 35(a)*), but can be built with a pitched roof to form a secret gutter (*Figure 35(b)*).

This type of wall is particularly exposed to the weather and it is essential that a high quality sulphate-free brick is used and a d.p.c. incorporated at roof level. The parapet wall to a pitched roof may be continued up the gables with the d.p.c. and copings. A cornice feature can be included as part of the parapet wall.

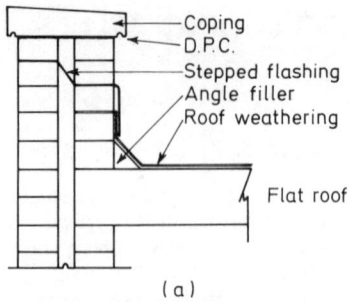

(a)

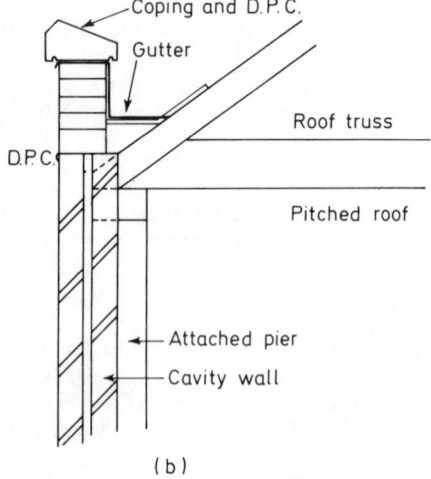

(b)

*Figure 35. Parapet walls. (a) With flat roof.
(b) With pitched roof.*

What are copings?

The feature at the top of a freestanding wall which gives weather protection and a decorative finish. The coping may be stone, concrete, brick or metal, and should offer protection and weathering to the top of the wall. Stone and concrete copings are very similar patterns; they have a weathered top surface, they project on both sides of the wall, have a throating under each projection and a d.p.c. to prevent moisture penetration through joints.

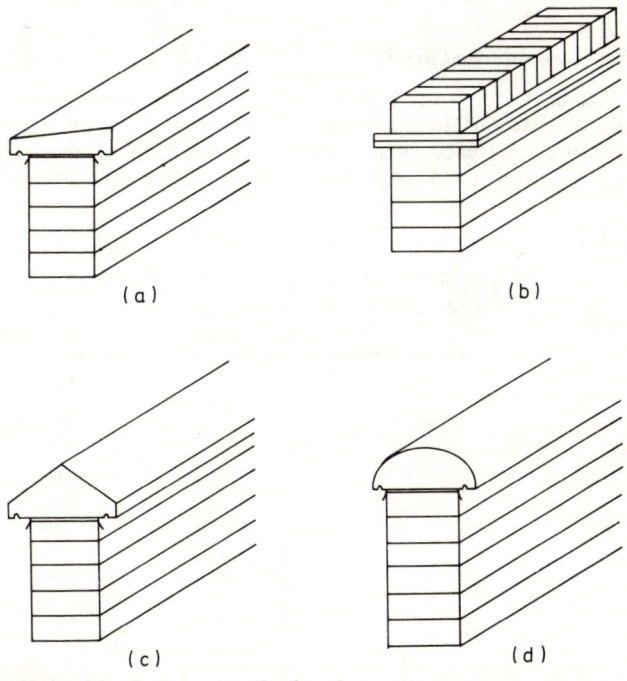

Figure 36. Copings. (a) Featheredge coping, stone or concrete. (b) Brick on edge, with two courses of tiles or slates. (c) Saddleback coping, stone or concrete. (d) Segmental coping, stone or concrete

Brick on edge copings have a weakness in that they have many exposed joints and do not project over the wall face. To improve this type of coping, two courses of slates or tiles are used under the brick on edge, to project over both faces of the wall and prevent the wall becoming saturated (*Figure 36(b)*). Engineering bricks are the best type for copings. Some purpose-made coping bricks are manufactured to special order but are rarely used in modern buildings. Metal copings are used mainly on parapets in conjunction with built-up felt roofing. They are usually of aluminium or copper.

What is an attached pier?

A pier projecting from and bonded into a wall. These piers are used at intervals to stiffen a wall and give it greater wind resistance. Attached piers may be on one side of the wall only or project on both sides. Their spacing and size are

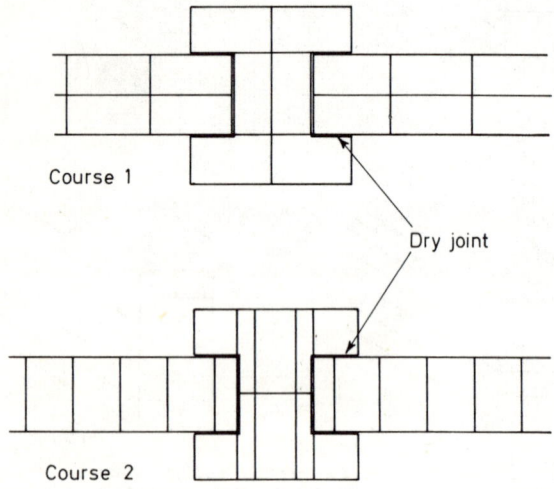

Figure 37. Movement joints concealed in an attached pier

determined by the height and thickness of the wall and its overall length.

Attached piers may also be used to increase the load bearing capacity of walls carrying roof trusses.

Movement joints are used in long walls to allow for thermal and moisture movements. It is possible to incorporate these joints in attached piers so that they do not show on the face of the wall (*Figure 37*).

How are attached piers bonded to the main wall?

The piers are bonded into the wall in the same way as a square junction would be bonded. If the projection of the pier is less than ¾-brick, then the bond on the face of the

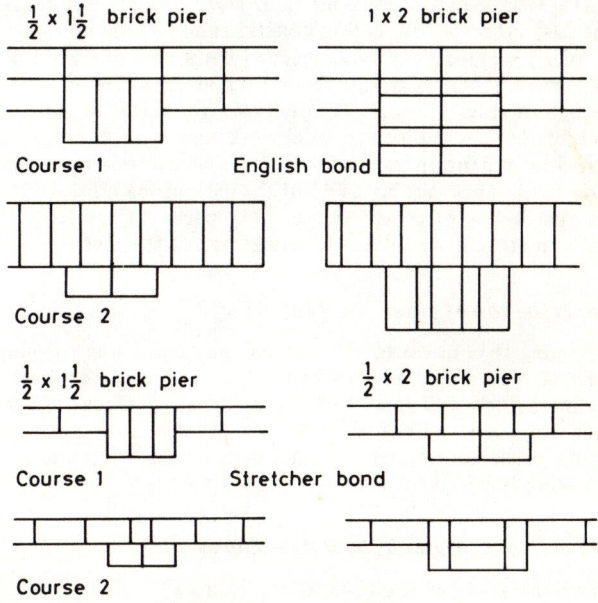

$\frac{1}{2}$ x $1\frac{1}{2}$ brick pier   1 x 2 brick pier

Course 1   English bond

Course 2

$\frac{1}{2}$ x $1\frac{1}{2}$ brick pier $\frac{1}{2}$ x 2 brick pier

Course 1   Stretcher bond

Course 2

Figure 38. Attached piers

59

pier is opposite to the bond in the same course of the wall. If the pier projects ¾-brick or over, then the face bond on the wall is continued through the face of the pier. Examples are given in *Figure 38*.

What forces are acting on a gate pier?

A gate imposes a considerable lever action on the pier from which it is hung, acting through the top hinge. The weight of the gate, also a person climbing over it, or a child swinging on it, and the force of the wind slamming it shut or open, have to be considered when building a gate pier.

The minimum size for a gate pier is 1½ bricks square, this is built as a Flemish bond unit and the centre either filled with concrete or a half-brick built in (*Figure 39(a)*). Reinforcing rods cast into the foundation concrete and passing up through the pier give greater stability needed for a pier of over 1.5 m high. Piers two bricks square or over are usually bonded in Flemish or English bond (*Figure 39(b), (c)* and *(d)*). If iron or steel hinges are built into brickwork and later corrode, the corrosion expansion can split the pier or cause it to expand. Galvanised types can be used but the galvanising tends to wear and rust takes place on the moving parts. Stains from the rusting metal can discolour the brickwork of the pier.

How is the top of a gate pier finished off?

A capping is built on to the pier and may consist of a stone or concrete slab weathered on top, projecting over the edge of the brickwork and incorporating a throating (*Figure 40(a)*). Brick on edge copings, with or without tile courses or projecting brick courses, give good weather protection and form a pleasing finish to the gate pier (*Figure 40(b)*).

How are boundary walls built on a sloping site?

Where the gradient is constant, the courses of brick or block-work in a freestanding wall are sometimes built parallel to the

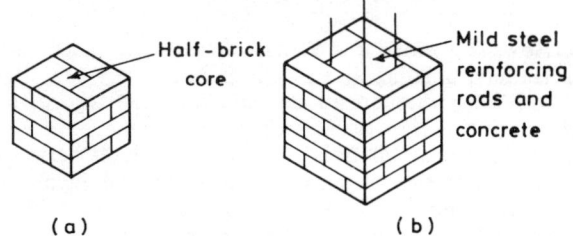

Isolated gate piers in stretcher bond

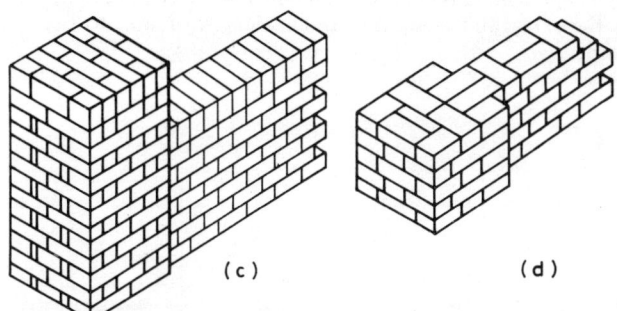

Figure 39. Gate piers. (a) 1½ brick pier. (b) Two-brick pier. (c) Two-brick English bond pier attached to one-brick boundary wall in English garden wall bond. Brick on edge capping to pier and coping to wall. (d) Two-brick Flemish bond pier with one-brick Flemish bond wall attached

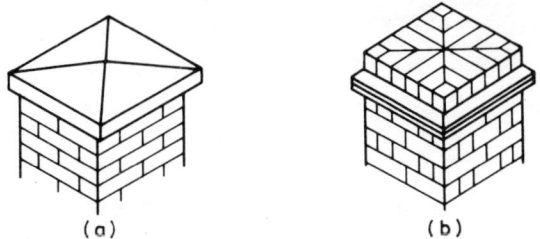

Figure 40. Pier cappings. (a) Stone or precast concrete. (b) Brick on edge with tile creasings

61

ground. However, where a wall has intermediate piers, gate piers or corners, building at an angle is not a very practical proposition. When the courses are built level the top of the wall can be cut to follow the gradient of the ground, or can be stepped up at regular intervals, the top being level between steps. A straight vertical step is the simplest method, involving little extra work, but alternatives are sometimes used, such as cutting the step at an angle or convex or concave curve to form a ramp. These methods are most suitable when a brick on edge type coping is being used, as the coping can be continuous up the ramp and along the straight length of wall to the next ramp.

6

OPENINGS IN WALLS

How are openings formed in brick walls?

Openings for doors and windows are formed using one of three methods. The opening may be set out by measurement and the jambs built up and plumbed, provision being made to fix the frame at a later stage. The accuracy of measurement and plumbing is of great importance if, subsequently, the frame is to be fitted properly. Another method widely used is to set the frames up into position and brace them securely, then build them *in situ*. This method is easier for the bricklayer but can lead to damage and distortion of the frame from falling material or careless working if the frame is fixed at this early stage of constructing the building. The third method is to use template frames which are set up and built into position as in the previous method, then removed to leave a true and accurate opening. The frame is fixed later when construction is more advanced. This method is generally used when hardwood polished frames are to be installed, as they can be stained by mortar splashes and require protection from construction damage.

What methods are used to construct the jambs of openings in walls?

The jambs are the sides of an opening, they can be square or rebated. The frame may be set in any position on a square jamb but is fixed into the rebate of a rebated jamb. The rebate gives extra stability to the frame and is generally used where very heavy doors and frames are to be fitted (*Figure 41(a)*).

Openings in cavity walls require vertical d.p.c. treatment at the jambs. Three methods of dampproofing the jambs of cavity walls are:

1. To close the cavity and build in a vertical d.p.c. up against the outer skin of the wall (*Figure 41(b)*).

2. To build up the cavity solid with slates and cement mortar (*Figure 41(c)*).

3. To leave the cavity open with no contact between inner and outer skins (*Figure 41(d)*).

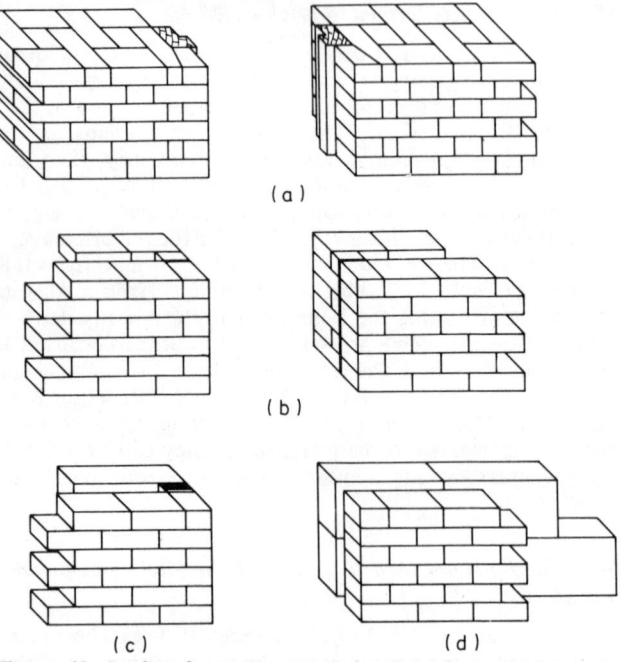

Figure 41. Jambs of openings. (a) Rebated jambs in 1½ brick Flemish bond wall – frame set into rebate. (b) Square jamb opening in cavity wall – closed cavity and vertical d.p.c. (c) Cavity closed with slates and cement. (d) Cavity left open

How are door and window frames fixed into openings?

Timber frames are fixed to the jambs by three types of fixing, *metal cramps*, *wood pads* and *plastics cavity closer*. The metal cramps are galvanised steel, one end bent and screwed to the back of the frame, the other fishtailed for building into the brickwork (*Figure 42(a)*). The frame is set up securely in position with cramps screwed on at the correct height to correspond with the bed joints of the brickwork. The first fixing is about 150 mm from the bottom of the frame and they are spaced at approximately 750 mm centres.

Wood pads (*Figure 42(b)*) are pieces of softwood 100 × 100 × 10 mm, soaked in a chemical preservative. The pads are built into the jambs by the bricklayer at positions similar to the metal cramps. The frame is nailed through into the pad by the joiner fixing the frame.

The plastics cavity closer (*Figure 42(c)*) is a preformed unit which has dovetail grooves to allow plastics cramps with fishtail ends to be inserted and built into the brickwork. The cavity closer is screwed to the back of the frame and built in by the bricklayers. A vertical d.p.c. is not required and the cavity wall is not closed with brickwork at the jamb.

A timber window frame is bedded on to a stone or concrete sill with a bedding mastic. Steel frames without wood sub-frames are bedded in mastic and screwed into fixing plugs cast into the concrete sill. Metal lugs bolted to the frames at the jambs are built into the brickwork in a similar way to cramps on wood frames. Timber door frames can be fixed at thresh-hold level by non-ferrous dowels fitted into the end of the frame and grouted into a hole in the top step.

What is the purpose of a window sill?

The sill is a dampproofing feature at the bottom of a window. The rain water which runs down the face of the window is made to run off to the external face of the wall and prevented from entering the building by the sill. The main features of an effective window sill are that:

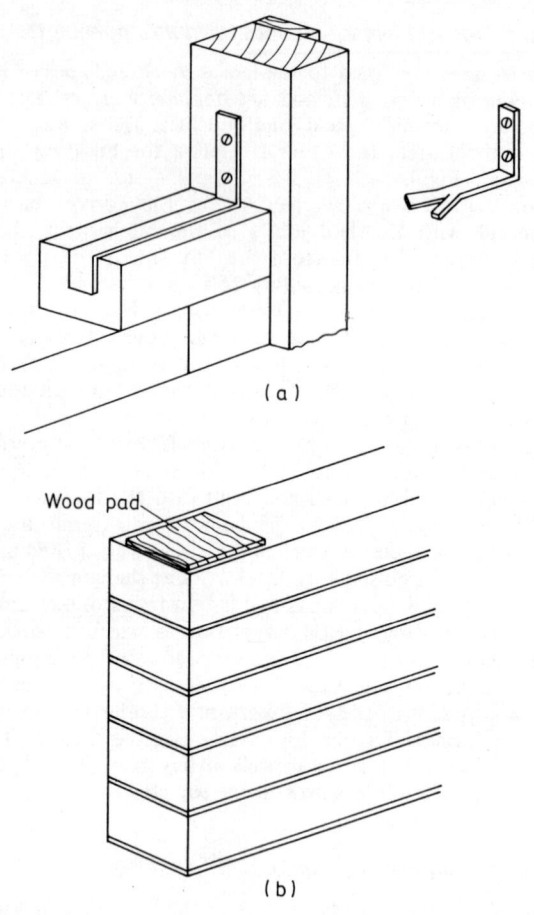

(a)

Wood pad

(b)

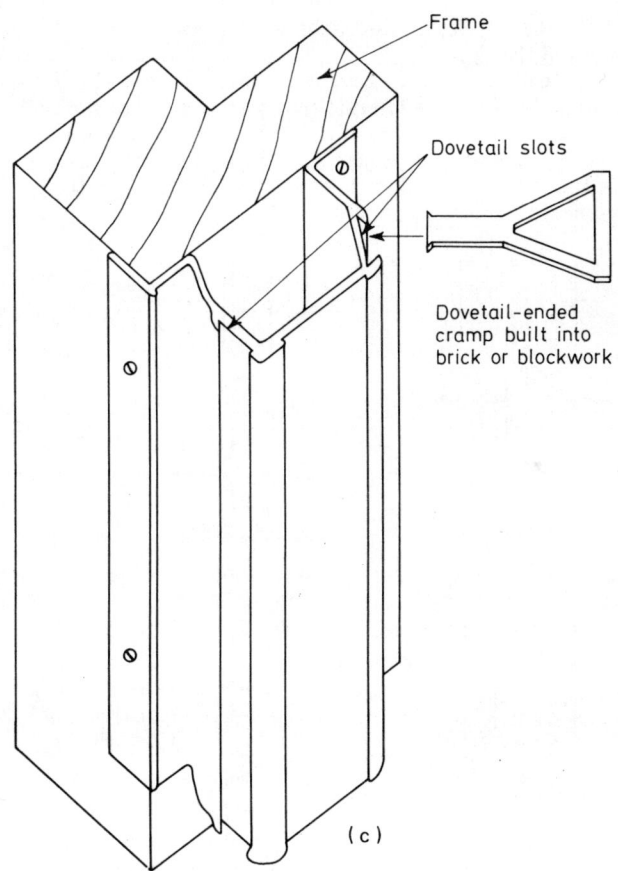

Frame

Dovetail slots

Dovetail-ended
cramp built into
brick or blockwork

(c)

*Figure 42. Frame fixings. (a) Metal cramps. (b) Wood pad.
(c) Plastics cavity closer*

1. It shall have a watertight joint between the window frame and the sill.

2. It shall have an effective d.p.c. incorporated where there is contact between the internal and external faces of the wall.

3. The top of the sill should be weathered to run water off effectively and should project in front of the wall face with a throating to ensure that rainwater drips clear of the wall underneath as much as possible.

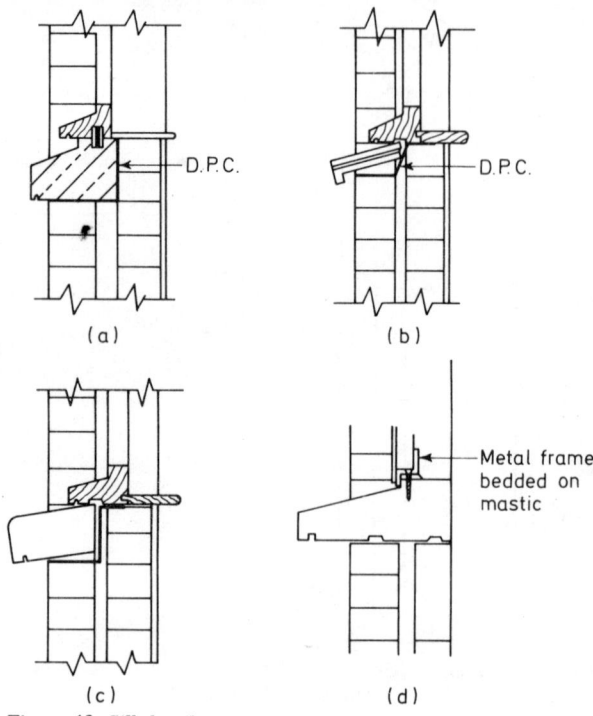

Figure 43. Sill details. (a) Stone or concrete sill with water bar. (b) Two-course tile sill. (c) Brick on edge sill. (d) Concrete and stone sill with metal frame fixed direct

Sills may be timber, sheet metal, such as galvanised steel or aluminium, stone, concrete, slates, tiles, or brick (see *Figure 43*). Built-up sills of slates, tiles (*Figure 43(b)*) or bricks have a potential source of weakness in that there are many mortar joints where water may seep through. Timber window frames usually have a hardwood sill which is bedded on to the structural sill. To prevent water penetration, a mastic joint is used, and with stone or concrete sills, a weather bar is incorporated. Steel window frames may be fixed directly to stone or concrete sills (*Figure 43(d)*) but require a timber frame for fitting to slates, tiles, or brick sills.

What is a threshold?

A threshold is the bottom of an external door opening and usually consists of one or more steps and a water stop to prevent rainwater being driven under the door into the room. The steps may be of concrete, stone or brick, the top step being built into the cavity wall. Methods of preventing rainwater penetration at threshold level include:

1. Fitting a water bar into the top step and using a rebated bottom rail on the door (*Figure 44(a)*).
2. Fitting an aluminium threshold which has a compressible p.v.c. insert in the top of it, the door being cut at an angle so that the p.v.c. insert is pressed tightly against the bottom of the door (*Figure 44(b)*).
3. Incorporating a gutter in the top step under the door. This gutter has a perforated grating over it and a drain-off pipe through the step (*Figure 44(a)*).
4. A hardwood timber threshold bedded on mastic and screwed down on the top step. This forms a water bar and as the door closes on to the top of it, there is clearance for carpet (*Figure 44(c)*).

What methods are used to bridge over openings in brick walls?

There are two methods used, *lintels* and *arches*. Lintels are beams of wood, steel, stone, concrete or brick. Built in above

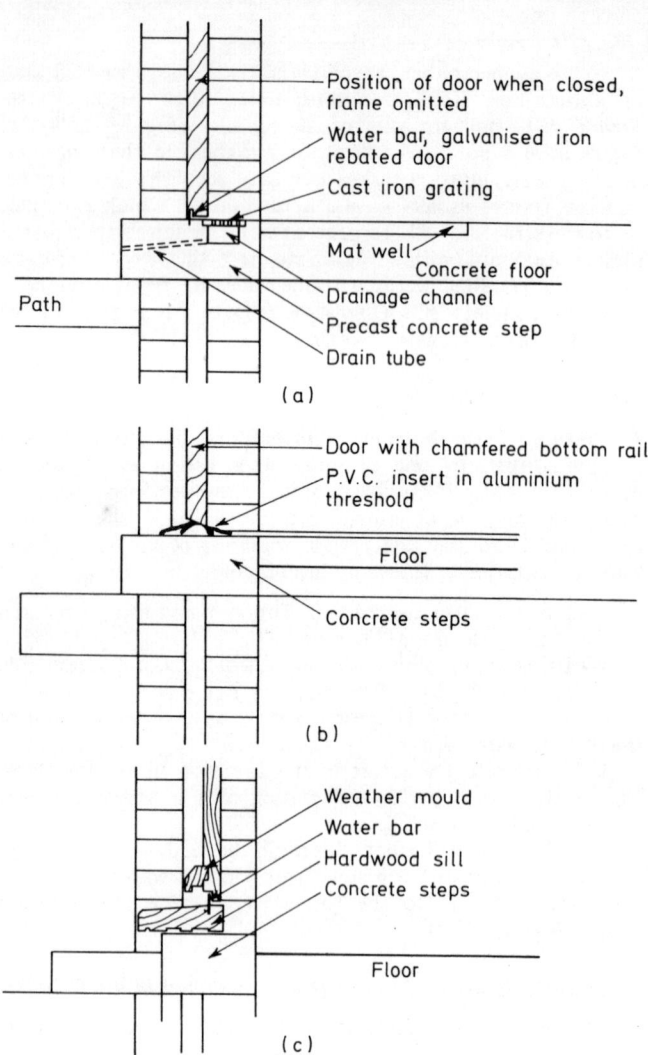

Figure 44. Thresholds

70

the opening to support the walling above, they present generally a flat, straight soffit for the frame to fix to. Arches are formed from wedge shaped units, usually of brick or stone, presenting a curved soffit for the frame.

What types of lintel are used in buildings?

British Standard beam sections are used, either singly or in combination, to span large openings, mainly in industrial buildings where large sliding doors or roller shutter doors are to be used.

Stone lintels are not used very much on modern construction because of cost. The span is restricted because, in many quarries, the natural length of stone is not sufficient to allow for wide span openings.

Concrete lintels (*Figure 45 (a)*) have largely replaced stone and are in very common use. They are reasonably cheap and when reinforced with mild steel rods, can be used to span the wide openings required in modern construction. The lintels may be rough finished to take plaster or rendering, or may be fairfaced to present a pleasing finished appearance. The section of a concrete lintel is generally rectangular but special shapes are cast. The *boot lintel* (*Figure 45(b)*) gets its name from the sectional shape. It is used to present a narrow concrete edge in the external facing brickwork and the main load bearing part of the lintel is on the inner face of the wall. The toe of the boot lintel may be projecting or set back from the wall face. The projecting toe is usually chamferred or weathered on top.

The BSC Ayrton lintel (*Figure 45(c)*) is a development of the Dorman Long type and has an angle welded to the back which makes a concrete support lintel unnecessary.

The corrugated galvanised steel lintel, *Figure 45(e)*, is used to carry partition blocks over openings up to 1 m span. It can be built into the normal bed joint thickness and provides a rigid lightweight lintel extensively used in domestic construction.

Dorman Long steel support lintels (*Figure 45(f)*) are galvanised bent sheet steel. The lintel rests on a concrete lintel

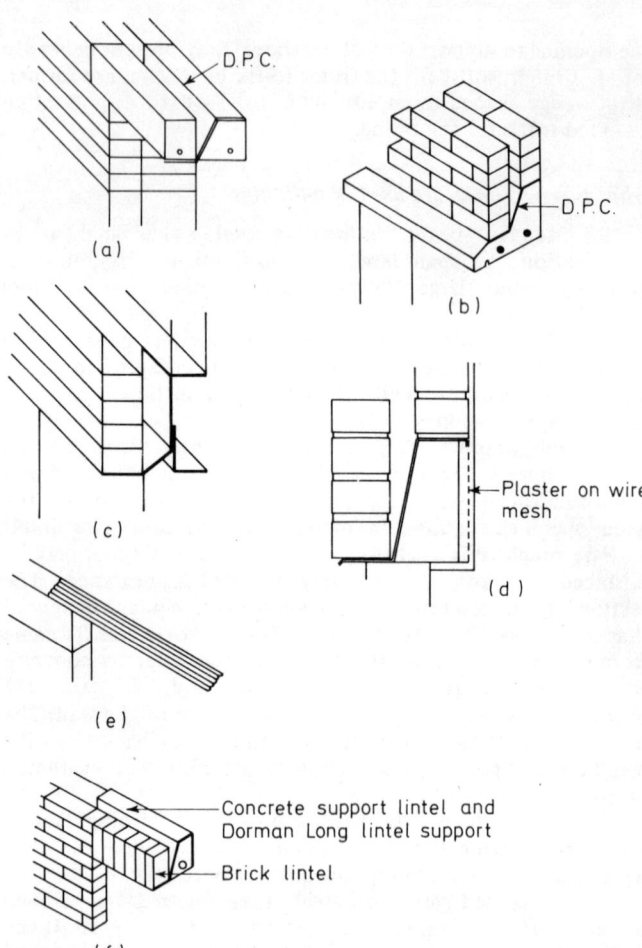

Figure 45. Lintels. (a) Concrete lintels. (b) Concrete boot lintel. (c) BSC Ayrton steel lintel. (d) Section of Catnic steel lintel. (e) Galvanised steel lintel for lightweight partition blocks. (f) Brick lintel

72

across the inner skin of the cavity wall. Brickwork is built across the toe of the steel lintel support, either as normal courses or as a soldier arch. The lintel support crosses the cavity and acts as a d.p.c. The Catnic steel lintel (*Figure 45(d)*) is similar but has a channel section welded to it which does away with the necessity to have a concrete lintel across the inner skin of the opening. Wire mesh is fitted on the inner face and underneath to allow for plastering.

Which timber is used for lintels?

Timber lintels generally were in very common use in the U.K., either separately or in combination with brick or stone, until concrete came into general use. The timber was usually oak, very often timber from broken up or wrecked ships. Nowadays, timber lintels are mainly softwood and their use is confined to internal partition walls in lightweight construction.

How is a brick lintel constructed?

The brick lintel, sometimes called a *soldier arch*, is not able to span openings without some form of support. This support may consist of a steel angle under, mild steel reinforcing rods and stirrups (*Figure 46(a)*), a sheet metal lintel support (*Figure 45(f)*), or tying to an *in situ* concrete backing lintel (*Figure 46(b)*).

The steel angle method requires additional temporary support during building in the form of a timber former on props and wedges, which is removed after seven days. The method using reinforcing rods requires substantial temporary support and is the most expensive, but if the opening being spanned has no frame, this method leaves the brick lintel with no visible means of support as the reinforcement steel is completely enclosed. The grouting must be carried out after the mortar jointing has set, so this method is the slowest. Grout is poured in from one end to ensure that the air is driven out and the completed lintel is solid. The galvanised sheet steel lintel support can be built on immediately after placing in position and requires no temporary support. This is the quickest and cheapest method. Brick or blockwork can be built across these lintels, giving the impression that no form of lintel is spanning the opening at all. This type of lintel support crosses the cavity and acts as a d.p.c. above the window. No damage can be caused when cleaning mortar droppings off the lintel.

The brick lintel with a concrete backing is another slow method, as the brickwork has to be built over a temporary support, formwork erected at the back of the brickwork and a reinforced concrete lintel cast *in situ*. The wall ties projecting from the back of the brick lintel are cast into the concrete to ensure that a strong tie is created between brick and concrete lintel.

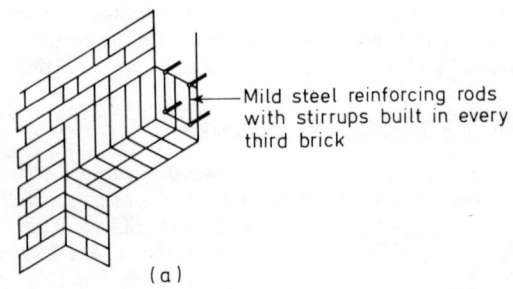

Mild steel reinforcing rods with stirrups built in every third brick

(a)

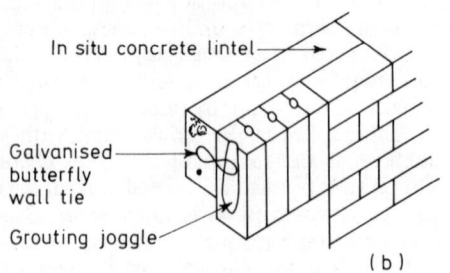

In situ concrete lintel

Galvanised butterfly wall tie

Grouting joggle

(b)

Figure 46. Brick lintels

Why is the reinforcing steel placed in the bottom of the concrete lintel?

A concrete lintel, when built into position, will tend to deflect slightly, causing stresses within the lintel. The top of the lintel will be subject to compressive stress, while the bottom will be subject to tensile stress. Concrete is a material which has great resistance to compression but is weak in tension. In order to prevent the lintel from cracking mild steel rods are introduced in the bottom to counteract the tensile stress. The steel bars are hooked or bent up at the ends and are placed so that at least 20 mm cover of concrete is ensured to prevent the steel from corroding and to give some protection in case of fire.

What is an arch?

An arch is comprised of wedge shaped units bonded together around a curve in such a way that the more the arch is loaded, the tighter it will bond together. Arches are of three main types:

1. *Rough arches* consist of uncut standard bricks with wedge shaped mortar joints. No cutting is required and this type of arch is used where a high standard of finish is not required.

2. *Axed arches.* The bricks forming the arch are all cut to a wedge shape to give a neat pleasing appearance. The mortar joints are even and parallel. This type of arch is used for facing brickwork, where a high quality of finish is required. Bricks can be purpose made for this type of arch to avoid cutting on site.

3. *Gauged arches.* The bricks are cut and rubbed to size with great accuracy and a very fine mortar joint is used. This type of work is very expensive and used only for fine quality ornamental work.

What are the parts which make up an arch?

The wedge shaped units making up an arch are known as *voussoirs*. The central unit of an arch, which is the highest part, is known as the *key brick*. This is usually the last to be built into the arch. The underside of the arch is known as the *soffit* and the line of the underside edge is known as the *intrados*; the upper edge of the arch is the *extrados*. The span of an arch is the distance between the reveals or jambs of the opening which the arch is bridging over, its rise is the vertical height from the springing line to the highest point of the soffit. Where an arch requires a sloping abutment for the springing this abutment is known as the *skewback*. The angle of the skewback varies with the span and rise of the arch.

How are arches supported during construction?

Where the arch has a small rise and span, a turning piece is used (*Figure 47(a)*). This is a solid piece of timber shaped to the curve of the arch which is set up and supported on props to the correct height.

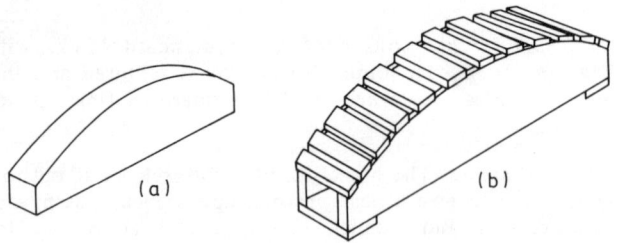

Figure 47. (a) Turning piece. (b) Segmental arch centre

Openings of large span and arch shapes of greater height cannot be built on a turning piece and so a centre is constructed (*Figure 47(b)*). The arch centre is made up from several timbers to form the shape required and can be used to span large openings. It is supported on props the same as a turning piece. In both cases the props have folding wedges at the top to facilitate fine adjustment and ease of striking and removal. Adjustable telescopic steel props can be used instead.

To enable the centre or turning piece to be removed when the arch is set it is usual for it to be shorter than the span of the arch. It should be remembered that timber expands when it gets wet and the centre could expand and tighten itself into the opening and have to be smashed out rather than removed carefully for re-use.

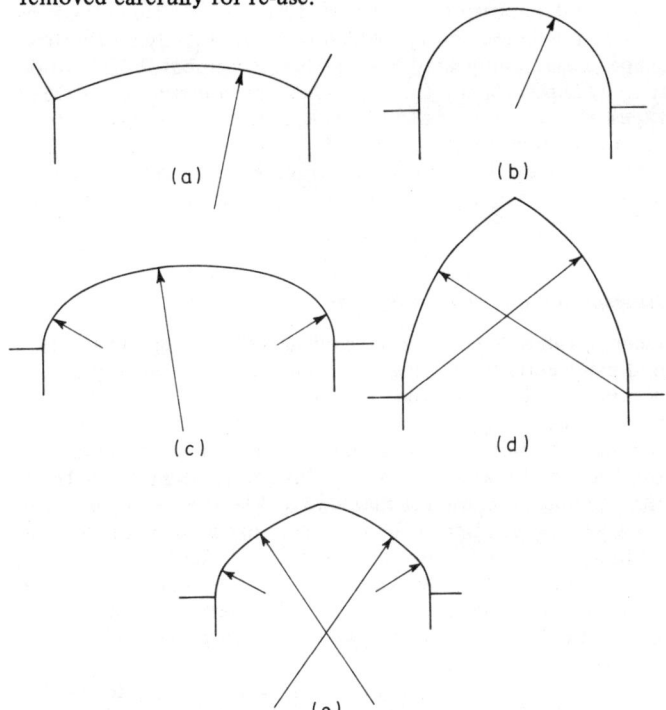

Figure 48. Arch forms. (a) Segmental. (b) Semicircular. (c) Three-centre. (d) Gothic. (e) Tudor

How many different arch shapes are there?

Although there are a great many arch shapes, they are all based on the segmental arch and the semicircular arch (*Figure 48(a)*) and *(b)*), which are the two most common types. The semicircular or roman arch has a semicircular outline, as the name suggests, and the rise is equal to half the span. The segmental arch is an arc of a segment of a circle and has a rise less than half the span.

A combination of arcs is used to produce a three- or five-centre arch (*Figure 48(c)*), which is approximately an elliptical shape. Less complicated are the pointed arches in the gothic style (*Figure 48(d)*). Gothic arches are constructed in three types: the *lancet* has a rise greater than the span, the *equilateral* has a rise equal to the span, and the *drop gothic* has a rise less than the span. The tudor arch (*Figure 48(e)*) also rises to a point, but not so pronounced as in the gothic arches.

How are temporary openings formed in walls?

Openings may have to be formed in walls during construction and made good later, although it is best to avoid this if possible. Making provision for the installation of large items of plant or equipment, or providing access to the interior of the building for construction purposes, may involve a large temporary opening in the wall. This would involve toothing the sides of the opening to allow for making good later and fixing a temporary lintel in place. A weak mortar mix is generally used for bedding and building over a temporary lintel in order that it may be removed easily when the opening is made good. The toothing must be built accurately and left clean, and when building in later, care must be taken to ensure that the toothing is packed solidly with mortar.

Smaller openings, perhaps for pipes, are usually formed by building bricks or blocks in sand at the required position. After the wall has set and hardened, the bricks set in sand are removed and an accurately formed hole remains.

How is an opening formed in an existing brick or block wall?

Frequently it is necessary to cut openings in existing walls for new doorways and windows. This is usually a fairly straight-forward piece of work, but it is essential to examine the wall thoroughly before starting. The examination should be carried out to establish the stability of the brick or blockwork and its general condition, the position of any cracks or any signs of movement, the condition of the mortar, and the position of other openings, roof and floor loads.

If the wall is in good condition and no faults are diagnosed, then no special precautions need be taken to support the building while work is going on. Existing openings in the wall above or alongside the position of the new one should be cross strutted between the jambs.

The first operation is to cut holes through the wall above the position of the new lintel. Needles, which are short lengths of timber or steel, are then passed through the holes and dead shores erected and tightened under each end of the needles. The spacing, positioning and dimensions of the needles vary with the width of opening and the particular circumstances of the job and can only be determined on site by an experienced person. The dead shores are either vertical timbers, standing on a soleplate and tightened in position by folding wedges, or adjustable steel props which are tightened by a screw jack action.

Having fixed the supports, the opening can now be cut into the wall and the jambs built up and made good as necessary, to support the new lintel. The lintel is then bedded into position and the work above made good up to the existing walling. It is essential that this making good is done very thoroughly and all joints packed solidly with mortar. After setting and hardening, the needles and shores are removed and the holes made good.

Small narrow openings may be made in walls of good condition without any temporary support. Floor joists supported by the wall above the proposed opening should always be supported by fixing a head plate under the joists and supporting by dead shores within the building. If the building has a timber ground floor, it may be necessary to cut holes in the floor and erect the dead shores on the site concrete, not on the timber floor.

7

FLUES AND CHIMNEYS

How is a fireplace opening constructed?

The hearth is a slab of concrete 125 mm thick, which extends from the back of the structural fireplace opening to project at least 500 mm in front of the opening. It must extend to a minimum of 150 mm either side of the opening. Where a solid concrete floor is used then the hearth is cast as part of the floor. In a hollow timber floor the hearth is supported by the fender wall in the case of a ground floor and projects from the chimney breast as a centilever in an upper floor. If the hearth is to serve a freestanding appliance not within a recess, then it must be not less than 840 mm square.

The chimney breast contains the flue from the fireplace and supports the fireplaces in the floor above, if any, and the flues and chimney. The standard fireplace opening measures 635 mm high, 355 mm deep from front to back and 580 mm wide, depending on the width of the appliance to be fitted. The lintel over the opening has a dual purpose of bridging the opening and forming the front of the throat. Standard lintels are often splayed on the back to assist in directing the smoke up the chimney or form a slab over the full depth of the opening and have a suitable hole in the centre to form the opening for the flue. The flue liners are bedded on to this slab and the flue outlet from the appliance fitted into the hole. The opening which has only a lintel over it has to be reduced to flue size by brick corbelling courses known as *gathering over*.

Fireplace openings can be constructed back to back in order to serve two rooms; if on the separating wall, to serve two houses. A fireplace situated on an internal wall will not lose heat by conduction through the back as it would if built on an external wall. The Building Regulations lay down minimum dimensions for the thickness of non-combustible wall at the back of the fireplace as follows. Where a fireplace is on an external wall or back to back with another opening, a minimum of 100 mm thick. A fireplace situated on an internal partition or back to back on a separating wall must have a minimum thickness of wall 200 mm up to a height of at least 300 mm above the top of the opening.

What is the fender wall?

A wall built to contain the supporting hardcore and carry the concrete hearth to a fireplace in a timber ground floor (*Figure 49*). The wall may also be combined with a sleeper wall

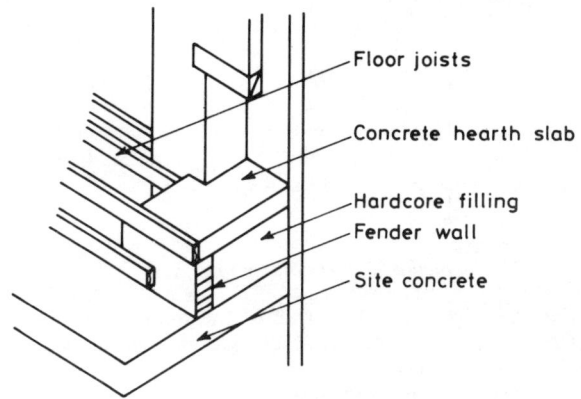

Figure 49. Fireplace opening

supporting the joists of the floor. A d.p.c. must be built into the wall at the same level as the other sleeper walls. A fender wall is not necessary to a fireplace in a building with a solid concrete ground floor.

How are brick domestic chimneys for solid fuel burning appliances constructed?

The commonest form of domestic chimney is a half-brick outer skin with a flue liner built to carry the flue gases. A single flue stack of this construction would be two bricks square outside size and built in stretcher bond. Stacks with two or more flues of similar construction are built in chimney bond which gives a Flemish bond appearance. The withes or mid feathers, which are the half-brick walls dividing the flues,

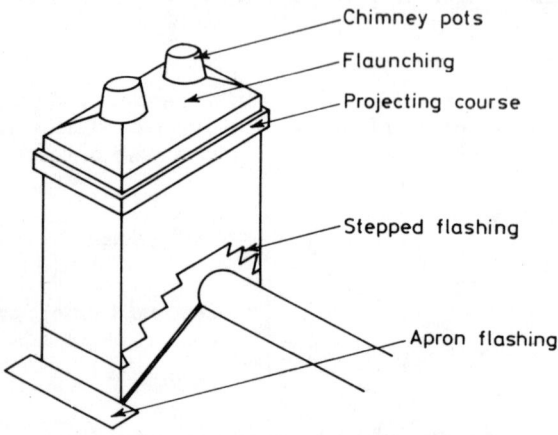

Figure 50. Two-flue chimney stack

are then tied into the outer wall of the stack. The flue liners are fireclay or refractory concrete sections, either square or round, which are rebated on the ends. They are jointed with a refractory cement and built as the chimney brickwork proceeds. Where a chimney passes through the roof space the brickwork must be rendered externally. A d.p.c. must be incorporated in the chimney where it emerges from the roof to prevent moisture seeping down through the brickwork into the roof space. This d.p.c. must be built in to the stack in conjunction with the flashings which prevent rainwater penetration between the chimney and the roof covering. The top of the stack has a chimney pot fixed to it which is a burnt clay flue terminal designed to ensure that the smoke and flue gases escape from the flue clear of the brick chimney. The chimney pot is firmly bedded down into the chimney which is then weathered with cement and sand flaunching to run the water off. To improve the weather resistance of a stack, which is usually in a very exposed position, a capping can be built on the top which consists of a projecting feature either in brick, concrete, or stone. This acts as a coping and makes the rainwater run clear of the top of the stack (*Figure 50*). Decorative features are sometimes incorporated into chimneys in the form of projecting or soldier courses. In some domestic designs, the chimney is deliberately emphasised and when built in artificial stone or contrasting facing bricks, becomes the outstanding feature of the building.

What is the minimum height above roof level for a chimney or flue pipe?

The top of a chimney, not including the pot, must be at least 1 m above the highest point of contact with the roof. If the chimney is within 600 mm of the ridges of a pitched roof then it may be 600 mm high. Where an opening window is within 2.3 m then the top of the chimney is not less than 1 m high.

What are the basic types of inset open fire solid fuel burning appliances?

The ordinary open fire grate is the simplest form of solid fuel burner. It is very inefficient, as most of the potential heat goes up the chimney and not into the room. Its efficiency and performance can be improved by installing a deep ashpit under the fire, to allow for removal of ash at intervals of a few days, or up to a week with some types. The ash pan, when full, is removed either by letting the fire out and lifting out or by removing from a door in an outside wall. Incorporated with the ashpit, these appliances have an underfloor supply of air through 100 mm pipes under the floor connected to ventilators in outside walls. The effect of this is to reduce the air being drawn into the fire from the room and to enable coke and similar fuels to be burned effectively. The air supply is controlled by a butterfly valve with controls set in the hearth. If the floor is a hollow timber one, the air can be drawn from the underfloor air space and air supply pipes will not be needed.

Incorporating a back boiler with this type of fire will improve its performance by providing domestic hot water. The back boiler replaces part of the fireback, and a flue, together with a control damper at throat level, is formed behind the boiler so that when the damper is open, heat is drawn round the boiler and maximum water heating is obtained. The open fire not only requires a great deal of air to support combustion but also draws air out of the room and up the chimney. The disadvantage is that draughts are caused in the room. To counteract this, the throat should be restricted to form a slot 300 mm wide and not more than 100 mm front to back – this is adequate for the smoke and restricts some of the excess air from entering the flue. Because the restricted throat makes chimney sweeping very difficult, an adjustable throat restrictor that can open to allow the passage of a flue brush can be fitted.

The open fire heats mainly by direct radiation, and to improve heating efficiency convection currents are used to circulate warm air round the room. The air is heated in a hollow chamber which surrounds the fire and convection takes the heat into the room. A cold air inlet is situated at hearth level and a warm air outlet above the fire opening. To further improve the efficiency of the open fire, a high output boiler may be included. This may take the form of a complete fireback boiler unit which surrounds the fire on three sides, or a specially constructed back boiler with flue incorporated in it. The high output boiler will provide sufficient hot water to heat a towel rail and radiators as well as domestic hot water.

How is a freestanding solid fuel appliance fixed?

The freestanding appliance may be in the form of a simple room heater which heats almost entirely by convected warm air, a boiler heating domestic water and supplying hot water to a radiator central heating system, or a combination cooker, boiler and room heater. Appliances of this type must be situated on a concrete hearth slab which extends 300 mm at front and 150 mm at the sides of the stove. The flue outlet from the appliance is connected into the flue liner and the joint packed with asbestos rope. Thermostatically controlled boilers allow very little heat into the flue and it is necessary to fit an access door and condensation tray at the bottom of the flue. A soot door into the flue is essential to allow for flue brushes where access through the appliance is not possible.

What are the two main causes of downdraught in a solid fuel burning appliance?

The downward wind current blowing on to the top of the chimney is often the cause of flue gases and smoke being blown back down the flue and billowing out into the room. This type of downdraught can be caused by the chimney being adjacent to high objects such as taller buildings, trees, and higher neighbouring chimneys. Another cause is the position of the chimney in relation to the surrounding land. The chimney may be situated in a depression in the ground or standing on sloping ground and the wind following the contours blows down on to it. Raising the chimney or fitting a taller pot may be the remedy in a few cases, but mainly this condition will require an anti-downdraught cowl fitted to the chimney outlet.

The second cause is differential air pressures set up by the wind flow around the building. The air pressure is usually positive on the windward side and negative on the leeward side. Negative air pressure will cause suction of air from a room through ventilators and round ill-fitting doors and windows. This air being sucked out of the room has to be replaced and may be drawn down the chimney. The process may be assisted by the fact that the chimney outlet is in a high pressure area caused by the wind, making a downward current of air in the flue. Downdraughts of this nature occur most frequently in bungalows and the top storey of blocks of flats, particularly where steep pitched roof slopes are involved, because the chimney outlet is in a high pressure area. Flues over 6 m high usually have sufficient movement of warm air to overcome the suction in the room but the air in the flue may lose velocity if the chimney is exposed and there is a rapid heat loss. Chimney pot designs vary and it may be that a tapered pot giving a venturi effect will increase the flow of air upward sufficiently to overcome air pressure at the top. Increasing the height of a chimney affected by air pressure may raise the outlet above the high pressure zone, but this will depend on the position of the chimney in relation to the roof. Ordinary open fires tend to suffer more than closed room heaters because they require a greater volume of air to flow through the throating and up the flue. This tends to cool the flue gases and thus reduce the velocity at the chimney top. It also means that where air is being sucked out of the room by negative air pressure, there is a conflict of demand and air is liable to be sucked down the chimney, bringing smoke with it. Louvred chimney pots tend to improve the flow of gases out of the flue by inducing a fierce updraught in the chimney pot when a wind is blowing.

What are the causes and effects of condensation in domestic flues?

The products of combustion of all fuels include water vapour. The amount of vapour released varies considerably and with solid fuel can vary with the amount of free water present. The water is evaporated when the fuel burns and the vapour is carried up the chimney with the warm air and the other chemical products of combustion.

If the vapour is cooled on its way up the chimney below dewpoint it will condense near the top of the flue. The condensate is contaminated with some of the combustion products, mainly oxides of sulphur. These oxides, when combined with water, form sulphurous and sulphuric acids, which have a corrosive action on many building materials. Cement and hydraulic lime mortars soften and often disintegrate completely from the effects of these acids.

Condensation attack usually produces the following results: leaning chimneys; cracked or split chimneys; loose brickwork; sand and dust at the bottom of the flue; brown stains and damp patches on internal chimney walls.

The factors which contribute to condensation are:

1. Burning wet fuel and wet rubbish.
2. Slow burning fuels which emit low temperature flue gases.
3. Slow burning closed appliances and boilers which produce low temperature gases and restrict the flow of air up the flue.
4. Flues too big for a modern slow burning appliance.

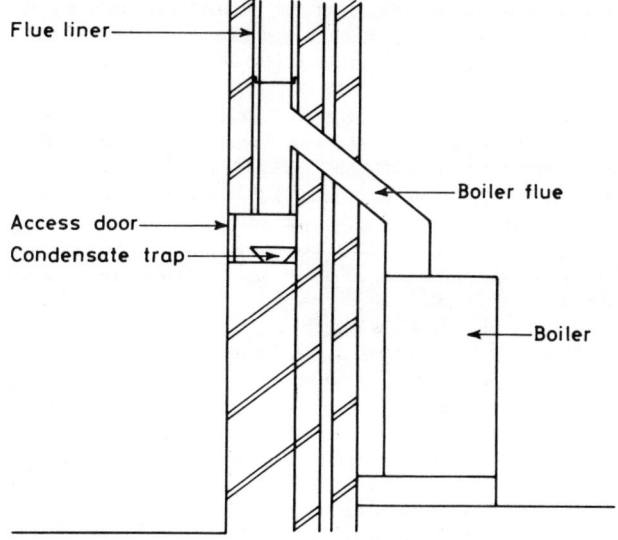

Figure 51. Condensate trap

5. Chimneys exposed on external walls which are kept cool by the prevailing wind.

The possibility of condensation must always be taken into account when designing and constructing flues. Careful attention should be paid to the construction of the flue liner, which should provide an insulated surface which will not reduce the flue gas temperature below its dewpoint. In very exposed situations, condensation may not be wholly prevented and at the bottom of the lining to an exposed external chimney there should be a condensate trap (*Figure 51*).

How is a flue liner fitted into an existing flue?

If the existing flue is straight then it is a relatively easy job to lower an asbestos cement liner down from the chimney after the pot and capping have been removed. This liner must be jointed to the flue outlet from the heating appliance or boiler and supported securely at this point. A concrete lintel with a hole formed in the centre is the best method of ensuring a good joint and supporting the weight of the flue pipe. If more than one length of pipe is used then a hole must be cut into the flue brickwork to allow for making the joint. Bends can be incorporated in this type of pipe but involve considerable extra work cutting out, jointing and making good. This tends to raise the cost of an otherwise fairly cheap material.

The flexible stainless steel liner, although much more expensive, can be threaded down from the top of the chimney or in some cases pulled up from the bottom. It can negotiate bends in the flue and can often be placed in position without having to cut into the flue at any point, thus saving expense on cutting out and making good. The liner is clamped at the top where a clamping plate is bedded at capping level. The light weight flue liner does not require support at the base and is directly connected into the flue outlet on the appliance, or into a rebated hole, as described for the asbestos pipe. The air space between the liner pipe and the surrounding flue brickwork is a good insulator, but if this space is filled with an insulating material such as vermiculite, then the insulation properties are greatly improved.

How are gas flues constructed?

Gas fired domestic appliances require a much smaller flue than solid fuel burners. Precast concrete gas flue blocks are made to bond into brickwork or blockwork walls, to construct the flue without any chimney breast projection. The blocks are of refractory concrete and have a rectangular slot in them, 300 × 50 mm or 375 × 50 mm, depending on the size of gas fire to be fitted. The blocks are jointed with a refractory mortar or a 1:3 lime—sand mortar, but are built into the walling with the standard mortar in use for that wall. A range of blocks is available, including fire recess blocks, lintels, straight and offset flues, front and side entry for boiler flues, and a variety of corbels and terminal blocks to suit all normal requirements. Double flue blocks are also available where more than one appliance is being installed.

The flue way, being a narrow slot, can easily be blocked during construction and it is essential that the bricklayer takes care to clear mortar squeezed out of the joint inside the flue. The flue terminal may take the form of a special ridge tile which has louvre openings in the side and is connected to the top of the flue. This is an unobtrusive terminal barely noticeable in the line of the ridge, and is usually an asbestos cement product. More conventional chimney terminals are constructed with brick, block or stonework cladding, usually with a cover slab and louvred flue exits in the side of the stack. Consideration should be given to the position of the terminal in relation to the prevailing wind in areas where persistent driving rain is common.

8

DECORATIVE FEATURES

What is a string course?

A string is a decorative horizontal feature course along the face of a building. It can be of brick, tile, stone or concrete, and frequently is used at sill or lintel height to form a continuous

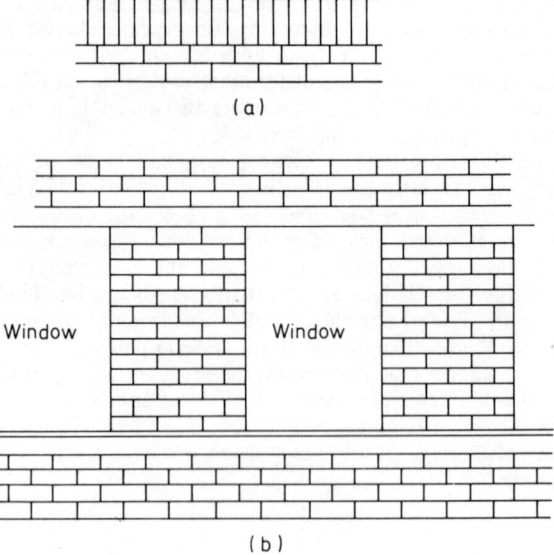

Figure 52. String courses. (a) Two courses of creasing tiles and a brick on end. (b) Continuous sill and head string

feature. A string course may also be used to separate different wall finishes, i.e. facing brickwork to first floor level and roughcast finish above. Some typical examples are shown in *Figure 52*.

How is the top of a buttress or attached pier weathered to ensure that rainwater runs off?

The top of the buttress must be weathered away from the wall by forming a sloping surface. This can be done by cutting the top course at an angle and bedding tiles or slates in cement mortar to give a waterproof finish.

Courses of plinth bricks can be used to reduce the buttress and to provide a weathered surface. *Tumbling in* is a method of building the top courses at an angle, continuing the face bond of the pier and tying in to the pier and the wall. The tumbling in courses are built to the angle by using a template or lines set up to the required angle; accurate gauging and setting out is required to ensure that the tumbling maintains the bond and ties in correctly at the top.

How are walls built, curved on plan?

Walls curved on plan are not built with special curved bricks unless a very small radius is required. Standard bricks can be used for curved work by taper cutting and by using V-shaped mortar joints. For radii up to 2.5 m header bond is preferred because the greater number of cross joints makes it easier for the bricklayer to form the curvature. Stretcher bond, or any bond which has mainly stretchers or complete courses of stretchers, is difficult on a lower radius because of the longer flat faces.

The curved wall is set out from a centre pin using a trammel rod and the wall checked with the rod as work proceeds. An alternative method is to use a curved template, but for a larger radius this becomes impracticable. At fixed intervals of about 1 m round the curve a lead brick is set, plumbed and gauged for each course, and then the work is built between the lead bricks and levelled between them. Each course is built in this way with the lead bricks acting as guides and reference points. A curved template may be used between leads on a large radius to help maintain the curve. A level datum point is required at each lead in order to enable the lead bricks to be gauged correctly.

What is a plinth?

A plinth is a projection at the base of a wall used where a tall or heavily loaded wall requires strengthening at the base by increasing the thickness. It is sometimes used in modern construction to give an illusion of strength to a building such as a bank or church. The plinth is reduced to main wall thickness by one or more courses of plinth bricks which have a chamfered face or end reducing the thickness of the wall by a quarter-brick. Plinth courses are difficult to bond in an English or Flemish bonded wall, and it is essential when setting out to consider the bonding of the main wall immediately above the plinth course and arrange the work below to suit this.

To avoid a straight joint on the corner immediately above the plinth it is sometimes necessary to use bevelled closers and to show what appears as a quarter-brick on the end of the wall for one course. In some cases, a double course of headers in English bond may be unavoidable but should if possible be confined to a relatively unimportant place.

What is a rusticated quoin?

Quoin is a term meaning the external corner of a building. The rusticated quoin is a decorative treatment of the corners to make them stand out and to act as a 'frame' for the building. Rustications take the form of projecting courses arranged in blocks, and may be built in a contrasting brick or incorporating different bonding or pointing to make them stand out from the main walls of the building. To be effective, rusticated quoins should be in scale with the rest of the building. Thus a bungalow may have rustications projecting 25 mm or less, three courses high and 1½ bricks long on the face. A very large building may have rustications projecting 50 mm, five courses high and three bricks long on the face. Rustications are frequently arranged to simulate the inband and outband quoin stones featured in masonry, by returning the projecting block half its length and building them alternately.

What is the difference between a corbel and an oversailing?

A *corbel* is a supporting feature projecting from the face of the wall. *Oversailing* is a course or a feature projecting from the general face of the wall.

Corbel and oversailing courses should project a maximum of a quarter-brick length. The total projection of a corbel must not exceed the thickness of the wall below the corbel. Corbel courses which project more than 25 mm should be built in header bond to gain greatest bond strength. Bricks with a frog in must be layed with the frog on the top to ensure even, solid bedding so that the corbel courses remain level. Corbelling and oversailing courses are generally lined along the lower arris, because this is the edge which is most likely to be seen, particularly as these features are frequently higher up on the building above eye level. A corbel is often used as a decorative and functional feature, as in a corbelled knee to a gable wall when the corbel may be built in tiles, bricks or stone. An oversailing course is often used as a drip course at the top of a chimney or as a string course feature.

How are decorative panels constructed?

When building a panel as a decorative feature, the surround is set out and built up to a suitable height, the panel work is set out on a flat surface and any cutting carried out before building. The panelwork is kept in line and plumb by lining from the surround brickwork. The panel may be of a special bond, typical bonds used as a decorative feature being basket weave bond (*Figure 53(a)*) and herring bone bond (*Figure*

53(b)). Natural stone or special facing bricks can be used as a feature panel and may be projecting from the face of the wall, or inset, which is the more usual form. The surround may be projecting and the corners mitred to increase the effect. Where the panel is inset on an external wall plinth, bricks may be used to improve the weathering at the base of the wall.

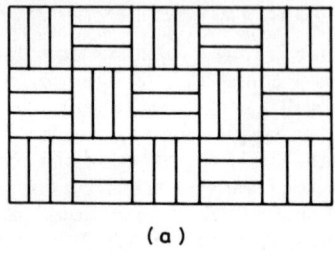

(a)

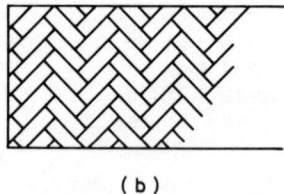

(b)

Figure 53. Decorative bonds. (a) Basket weave
bond. (b) Herring bone bond.

9

PARTITION WALLS

What is the difference between partition walls, party walls, and compartment walls?

Partition walls divide a building up into rooms and corridors. The partitions may be a load bearing part of the structure or non-load-bearing, they may be permanent or temporary, demountable types, which can be moved readily to give a different room layout within a building. The party wall is a wall which divides two adjoining properties and which is common to both owners.

Compartment walls divide the building into fire resistant compartments. The walls must have fire resistance as shown in the Building Regulations and any permitted openings for ducts, pipes and chimneys, etc., must be adequately protected against passage of fire.

What are partition walls?

Partition walls divide the floor space in the building into rooms and corridors. They may be simple dividing walls, non-load-bearing, or may be a load bearing type designed to withstand structural loads as well as divide space.

Brickwork and blockwork partitions are often preferred to cheaper forms of sheet construction because they offer greater resistance to the passage of sound through a building.

What materials are used by the bricklayer to construct partition walls?

Brickwork has largely been replaced by blockwork for partition walling, mainly for reasons of economy and weight reduction.

Non-load-bearing partitions which are simply space dividers are constructed from lightweight non-load-bearing blocks, usually of the hollow type. Partitions of this type are easy to build, relatively cheap, and do not impose much loading on to the main structure of the building. They are usually plastered both sides but fair faced blockwork is frequently used to good effect, either plain or decorated by painting. Load bearing partitions require heavier construction and blockwork must be able to resist the loads imposed — dense aggregate blocks and solid lightweight blocks are used. Brickwork is used for partitions where facing brickwork is required for architectural effect or where the loads imposed require the higher crushing strengths of calculated brickwork.

How is a partition wall stabilised during and after construction?

Brick partitions of half-brick thickness and block partitions over 100 mm thick or more can usually be built up to scaffold height and up to storey height without much difficulty. Scaffolding must be independent and particular care must be taken with short, straight lengths of wall which may be easily overturned.

Thin wall partitions of blocks down to 65 mm thickness and of lightweight non-load-bearing blocks may be very unstable during construction and may collapse under their own weight as work proceeds. It is necessary to erect profiles of a temporary nature in order to build partitions of this type. These profiles consist of 100 × 50 mm timbers which have at least one straight edge. These are wedged into position between floor and ceiling at corners or stop ends and at central positions on longer wall lengths. Door or window openings are measured and either the permanent frames or temporary profile frames set up in position. These aids enable the bricklayer to erect the wall with the minimum of difficulty. The gauge for the work can be marked on the profiles and the building line raised on the profiles instead of on the blockwork. Partitions are relatively thin for their height, and as such are easily overturned by the wind. Timber supports may be necessary to give added stability to a completed wall until the building has reached a more advanced stage of construction and the wind has been excluded.

A non-load-bearing partition may be taken up to the soffit of a concrete floor or beam and cement mortar packed tightly into the joint at the top, but there will be some drying shrinkage on the blockwork and a crack will inevitably occur at the joint. A better method, particularly with a frames structure where slight movement can also be expected in the frame, is to leave a dry joint or build in a strip of fibre board. The joint can then be masked with a suitable cover mould.

Where a false suspended ceiling is to be installed, the partition acts as a barrier to sound transmission through the ceiling by breaking the ceiling up into room size panels. It also prevents the rapid spread of flame and acts as a fire stop. The false ceiling work acts as a stabilising restraint at the top of the partition, preventing it from overturning. Short lengths of non-load-bearing partition are usually sufficiently stabilised by tying in securely at each end, either to other walls or by fixings to columns.

What are storey height frames and what is their purpose?

Door and window frames which rise the full height of a storey in a building are known as *storey height frames*. They are fixed at floor level and at ceiling level, adding greatly to the stability of a non-load-bearing partition. They effectively divide a partition into smaller isolated units, which eliminates the possibility of shrinkage cracking occurring above a door opening of conventional height. The space above the door may be infilled with a solid panel or glazed to act as a borrowed light.

10

CLADDING PANELS

What is meant by natural stone cladding?

A thin facing of natural stone slabs fixed to the face of a building or wall to achieve architectural effect. The cladding does not carry structural loads but is itself supported by the structure. The cladding must be fixed in such a way that it can move with a slight degree of independence from the main structure, to allow for variable thermal movements. Cladding must not be confused with natural stone tiling.

What is precast concrete cladding?

Facing slabs of precast concrete can be used as a cladding in the same way as natural stone. The thickness of the slab varies from 50 mm to 100 mm depending on the surface treatment and the area of each individual slab.

The concrete slabs may be reconstructed stone, which is a surface made to resemble a natural stone with a concrete backing. The stone face is obtained by using natural stonedust as an aggregate and cement of a suitable colour. Cast stone slabs of this type can be of plain surface finish or given a 'rock' face or a profiled face appearance which will give changing light and shade patterns as the sun's position changes.

The texture of the surface of the concrete cladding can be made interesting by exposing the aggregate. This can be done by grinding off the slab surface or by rolling a selected aggregate into the slab surface immediately after casting. The most widely used method is to incorporate the aggregate into the slab and wash off the surface cement and sand, leaving the aggregate exposed. Large pieces of special aggregate can be laid face down in the bottom of the mould on a 10 mm sand bed and the concrete in a semi-dry state laid and compacted on top of it to make up the thickness of the slab. The sand is washed off the face after the slab has hardened and been removed from the mould.

A great variety of surface finishes can be obtained by using colours, textures and shapes which are unobtainable in natural stone.

What types of stone are used for cladding?

Good quality limestones and sandstones, free from any defects and of a durability suitable to the situation in which they are to be used. The recommended thickness for slabs of these stones is 100 mm. The area of each stone slab should not exceed 0.75 m². On no account should limestone and sandstone be used together, as the soluble sulphates in the limestone attack and break down most sandstones.

Granite, marble and slate can be used. The recommended thickness for external claddings in these materials is at least 37 mm, though certain types of stone will need to be 50 mm thick because of the need to ensure a minimum fixing and cramping strength. Internal claddings in these materials should not be less than 20 mm thick. Quarry variations, particularly in slate, mean that sometimes a thin slab has to be thickened to provide cramping points. This is done by fixing thickening pieces with adhesives and pins. Slab sizes vary, but generally material 20 mm thick should not exceed 0.5 m² and material 37 mm thick should not exceed 0.9 m² in area.

Any cramp holes or slots should always be drilled or sawn in the stones to avoid stunning or fracture around the edge of the hole. Because stone is a variable natural material, the suppliers' recommendations should always be followed.

How are stone slabs fixed to the face of a building?

Cladding slabs are fixed to the face of a wall by means of metal cramps. These grip the stone and are built into the backing wall or fixed into concrete backing. Cramps should be non-ferrous metals — steel or iron should never be used. Metals in general use include copper, phosphor bronze, gun metal, and in certain cases stainless steel alloys. Because of the possibility of electrolytic corrosion, metals should not be mixed in the same job.

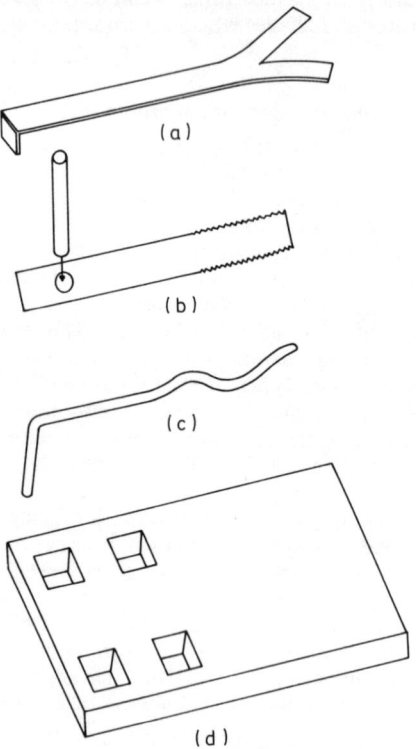

Figure 54. Fixing cramps. (a) Flat fishtailed. (b) Dowel fixing with serrated end. (c) Bent wire cramp. (d) Flat corbel plate

Various types of cramp are manufactured (see *Figure 54(a)*, *(b)* and *(c)*). Generally, round wire cramps are used in the fixing of thin claddings such as granite, marble and slate, while flat sections are used in the fixing of the heavier limestone or sandstone slabs. Cramps intended for building in to brick backing walls have fishtail or serrated ends, while a dovetail ended cramp is used in conjunction with a cast *in situ* anchor slot for fixings into concrete units.

How is the weight of stone cladding supported?

Stone cladding has to be supported by the backing wall or by the frame of the building to which it is fixed. In the case of a building, a system of supporting the cladding must be used at every storey height.

One method is the bonder course frequently used with a framed building. The bonder is a course of cladding which is bedded into the backing wall and acts as a corbel supporting the cladding above it, transmitting the load to the structural wall or frame. The bonder may be the same size as the other cladding slabs on the face or may be used as a feature course similar to a string.

To make sure that the bonder course fully transmits the load to the structure, a compression bed is built immediately under to prevent loads being transferred to the cladding below. A bonder course of the wrap-over type often forms the beam cladding at the head of a window opening below. A reinforced concrete boot lintel or a nib cast on to the face of the concrete can be used as a cladding support for the thicker slabs (100 mm). A notch is cut out of the back of the slab to correspond with the nib and half the thickness of the slab, which is then bedded on with mortar. It is also possible with some types of stone to have a horizontal chase cast into concrete and a corresponding projecting nib formed on the back of the stone.

Non-ferrous metal corbel plates can be used, again at each storey height (*Figure 54(d)*). The plates are built into the brick backing or grouted into the concrete in such a position that they coincide with a vertical joint and thus support two slabs. The cladding has slots cut into the back which fit the corbel plates. Alternatively, angled corbel plates can be fixed to concrete by means of non-ferrous expander bolt fixings. Obviously great accuracy is required when using corbel plates and very little tolerance is possible. A non-ferrous metal angle fixed to the concrete frame by means of non-ferrous expander bolts is sometimes used as an alternative to corbel plates, but is more expensive. It may, however, form a feature strip at the head of a window opening.

Why are expansion and compression joints necessary in cladding?

Because the cladding is a skin on the outside of the structure it is more affected by temperature changes, and provision must be made for differential thermal movement between the cladding and the structure as well as in the cladding itself.

Vertical expansion joints must be incorporated in any cladding. The distance between these joints will vary because of the design of the building, the area involved and according to whether joints are incorporated in the main structure. Usually, concrete, slate, granite and sandstone require joints at about 7 m intervals, while marble and limestone require a spacing of about 10 m. Where areas of cladding are restrained at the sides by vertical structural members, any expansion joint is advisable between the cladding and the restraining feature.

Expansion joints can be filled with a joint sealer having elastic and adhesive properties not affected by temperature or weather, and which will retain these properties indefinitely. Bituminised foamed polyurethane or polysulphide sealers are very suitable for this purpose. The former is a preformed strip type of joint, the latter a soft mastic which is applied by means of a hand operated pressure gun.

Spring copper can be used in a slot cut in the edges of the slabs. This gives an external appearance of a dry open joint. In some cases a sliding joint is incorporated at the junction of two wings of a building or at an internal or external angle.

At an internal angle, one face of the cladding passes behind the other face for the full height and the straight vertical joint is sealed. On an external angle, the corner may be bonded in the traditional way, but quoin stones are set in soft lime mortar with mastic vertical joints.

Compression beds are located under a bonder course or under a support course. The object is to prevent the loading on the cladding becoming excessive and causing disintegration. The joint sealers used include butyl rubber tapes, and bituminised foamed polyurethane, both built in as the work proceeds, and polysulphide mastics applied by gun. Expansion and compression joints are similar width to the normal slab joints, i.e. 5 mm.

11

DRAINAGE WORK

What are the main requirements of a drainage system?

Drains should be self-cleansing and should operate without maintenance. The pipes should be strong enough to resist the pressures placed upon them in the ground. The bore should remain true and smooth to ensure that the water flows as freely as possible, and the gradient should be such that a water velocity is maintained which will ensure that any solid matter will be floated along the pipe. Pipes should be laid in straight lines, between points of access whenever possible. Ventilation of the drainage system should be maintained in order to prevent a build up of gases. All entrances into the drains should be trapped to prevent foul air from escaping.

What types of pipe are used for underground drainage?

Pipes can be divided into two types, rigid pipes and flexible pipes.

Rigid pipes are glazed stoneware, asbestos cement, concrete and cast iron. They may have rigid or flexible jointing systems. The flexible pipes are pitch fibre and unplasticised p.v.c.

Flexible drainage systems offer advantages in that they can accommodate ground movements without fracture. Flexible pipes, while offering advantages such as ease of laying, lightness, long lengths and fewer joints, have been found to become compressed by the weight of backfill on top of them and in some cases to have flattened. Rigid pipes will resist this type of pressure, but can fracture if the bedding offers uneven bearing and have been cracked by careless backfilling.

How are the different types of pipe jointed?

Clay pipes are manufactured with and without sockets.

Socketed pipes can be jointed by inserting the spigot into the socket, caulking with tarred hemp and then making the joint with 1:1 cement and sand mortar. This rigid joint is comparatively slow to make, is affected by the weather and cannot be tested for at least 12 h. Flexible joints which are used with socketed pipes consist of a rubber O-ring which is compressed between the spigot and socket, or preformed polyurethane gaskets which are pushed together. The plain ended clay pipe is jointed by fitting a polypropylene coupling with rubber sealing rings which tighten on the pipe when the coupling is pressed on.

Cast iron pipes are rigidly jointed by running and caulking a spigot and socket joint with lead. Rubber ring or glands are used to provide flexible joints.

Concrete pipes are jointed in the same way as clay pipes but concrete pipes without sockets have rebated ends and do not have waterproof jointings.

P.V.C. pipe jointing systems consist of preformed sockets or loose couplings, and the joint is made either by a rubber O-ring or by using a solvent as an adhesive. Asbestos cement pipes have tapered ends which are jointed by means of a coupling and rubber rings. These pipes are mainly used for drains which are carrying water under pressure from sewage pumping systems. Where a cut length has to be used, a new tapered end has to be worked on to the pipe to take a coupling.

Pitch fibre pipes are jointed by means of external couplings in polypropylene. One type is a tapered coupling which is driven on to a pipe with a similarly tapered end. The other type of coupling fits over a plain ended pipe and incorporates a rubber sealing ring.

All pipe systems can be interconnected by using specially made adaptors.

How are drains tested?

The testing of drains is the responsibility of the local authority in whose area the drainage work has been carried out. The

Building Regulations require that all drainage work be tested for watertightness upon completion, and tests of water, air and smoke pressure are usually applied at the discretion of the local authorities inspector.

The water test consists of filling the pipe run under test with water and maintaining a head at the top of 1.2 m and not more than 2.4 m above the lowest point. In order to be able to fill the drain, an expanding plug must be fitted at the lower end. The most convenient place to fix this is in a manhole. The internal pressure placed on the pipe by this test is similar to pressure that would be applied if the drain was blocked and filled up during use. The test should be maintained for one hour. Some topping up may be necessary to make allowance for initial absorption.

The air test is easier to apply, but if the drain has a small leak it is sometimes difficult to find. To carry out this test, the drain length must be plugged at each end and an air pump with pressure gauge or manometer attached to one of the plugs. Air pressure is built up in the drain by means of the pump until a head of 100 mm is registered by the manometer or pressure gauge. This pressure should not fall below 75 mm during a five minute test period.

Smoke testing is carried out by placing a smoke cartridge in the lower end of the drain and fitting a drain plug. The head of the drain is left open until smoke appears and then plugged. Traces of smoke escaping will indicate any faults in the drain. This test can be carried out using a smoke generating machine which introduces smoke into the drain under pressure. A floating copper dome rises to the top of the machine and should keep this position for at least three minutes. Leaks in the drain will be indicated by the copper dome settling down into the machine.

When carrying out air or smoke tests on drains with trapped gullies at the head, a rubber tube must be pushed through the water seal to prevent an air lock developing in the drain and possibly bypassing the test.

An electric torch and mirror are sometimes used to inspect the bore of a pipeline after it has been laid. The mirror is set at an angle at one end of the drain run and the torch is placed at the other. The light from the torch will be reflected in the mirror at the other end, and if the bore is true a circular light image will appear.

How does the method of sewage disposal influence the type of drain and sewer system?

Foul sewage can be discharged by means of an outfall pipe into the sea. This system is relatively cheap because the sewage does not need any form of treatment, but it is being severely criticised because of the sea pollution that results. A more expensive method is to construct sewage treatment works, in which the sewage passes through a series of settlement tanks and filter beds which clean the water of all organic impurities and discharge relatively clean water into rivers, lakes or the sea without any pollution resulting.

Where such a sewage treatment works exists it would be uneconomic to pass all surface water through the works, as this can be discharged direct into natural water courses. In this situation the ideal is a separate system of drainage. All foul water is carried in the foul drain and sewer to the treatment works. All surface water is connected into the surface water sewer and discharged into natural water courses.

This system is costly to instal, as it requires two sewers and two drainage systems. As an economy measure, a partially separate system is operated by many local authorities. Under the partial system, as much surface water as possible is discharged into soakaways, natural water courses, lakes, sea, etc., and only one sewer is provided. Where it is unavoidable, surface water is connected into this single drain and sewer. The combined system operates mainly where there is a sea outfall and no treatment works. Only one drain and sewer system is required and all drainage is carried in this.

Mainly, the sewer is the responsibility of the local authority, while the drain is the responsibility of the owner of the property which it serves.

What is subsoil drainage?

Subsoil or field drainage is a system of pipework in the ground which collects ground water and discharges it into surface water sewers or into natural drainage watercourses. These drains are used to reduce surface flooding and thus improve the stability of the ground surface and in certain conditions to reduce the water table. Drains of this type collect surface water which is on the ground surface and ground water which is in the ground above the natural water table level of the subsoil water.

Pipes used for this type of drainage are either butt-ended porous pipes of unglazed clayware or no fines concrete, or perforated pipes of clayware, concrete, pitch fibre or polythene. The pipes are laid at the bottom of a trench and a surround of broken stone placed around the pipe. This material may also be used to backfill the trench to the top to facilitate entry of surface water. These are known as *French drains* and are sometimes constructed as a narrow trench filled with broken stone rubble and without a pipe.

To collect the water flow from several subsoil drains and to provide access for removing silt, a brick or concrete catchpit or siltpit is constructed. This is a brick walled box with a concrete base and cover. The inlet pipe or pipes are set 25 mm above the outlet, and the base of the pit is a minimum of 600 mm below the outlet. Water flows in and any silt and dirt settles to the bottom of the box.

Catchpits are frequently constructed where a branch drain connects with a surface water sewer.

What is a soakaway?

A soakaway is used to disperse surface water which cannot be run into some natural watercourse. The soakaway is a pit dug in a suitable position in permeable strata. This pit receives the discharge from a surface water drain or drains and allows the water to seep away into the ground. Generally, the soak-away is filled with large broken stones and has a slab over to exclude soil and to support the earth over the soakaway.

Some soakaways are lined with brick or stone walling constructed with open joints or in concrete with perforated sides. These are not filled with broken stone and therefore have a greater capacity. It is important that the soakaway is placed in a position where the water will percolate away so that it will not be permanently waterlogged. Perforated precast concrete soakaways are stood on a bed of hardcore and surrounded with similar material, and in certain types have an access cover fitted to allow for inspection and possible cleaning.

What provision is made for inspection, maintenance and rodding of drains?

Access to the drainage system is through access fittings, rodding eyes or inspection chambers.

Access fittings are drain fittings such as gullies which carry out their normal function but are provided with an access point to allow for rodding, which has a cover that can be bolted tightly to a rubber gasket.

Rodding eyes are usually constructed at the top of a shallow drain. A bend is fitted to bring the end of the pipe near to the surface of the ground, a stopper is set in grease in the bend and a cast iron access cover placed over it.

Inspection chambers or *manholes* are constructed of brick, concrete or g.r.p. and concrete. They consist of a concrete base with the drain in the form of a half-section channel and branches of any flowing into it. This is enclosed in a box of brick or concrete with a concrete cover and a cast iron access cover. Inspection chambers may only be 0.6 × 0.45 m inside dimensions on shallow domestic drainage systems, but can be very large structures on deep sewers.

A general principle is that all lengths of drain or sewer should be accessible for testing. Inspection chambers are recommended in the following positions: at junctions, between a branch and a main; at changes of direction of the drain, i.e. where a drain has to be taken round a building — minor

bends of less than 45° do not require access; at the point of
connection between a drain and a sewer or within 12 m of the
connection; at changes of gradient.

How are inspection chambers constructed?

Inspection chambers are boxes constructed in the ground of
brickwork, *in situ* concrete, precast concrete, or plastic. In
each case a concrete base slab of 150 mm is laid in the bottom
of the excavation. The top of this concrete is below the
required invert level of the pipe and allowance must be made
for the pipe thickness.

Brick construction. The channel's branch bends and con-
necting pipes are laid and jointed in position and the first
courses of brickwork are cut round these pipes. If the diameter
of the main drain or any of the branches is 300 mm or more,
an arch is built over the pipe. Class B engineering bricks, 1:3
cement-mortar, and English bond should be used and care
taken to bed and joint the bricks solidly to ensure a watertight
construction. The inside of a chamber is plumbed and flush
pointed as this is the face side. A reinforced concrete cover
cast *in situ* or precast has an opening in it of suitable size for
the metal access cover to be bedded on top of two courses of
brickwork.

In situ concrete construction. Formwork is erected and the
concrete of 1:2:4 mix is well rammed into place. The concrete
cover and metal access cover are fixed as for brick chamber.

Precast concrete. Usually circular, the chamber is constructed of a number of units joined together until the required
height is reached. Reducer slabs and taper pipes are used to
reduce the chamber to access cover size and the metal access
cover bedded as before.

G.R.P. manholes are preformed plastic manufactured in
several standard branch layouts and heights. They are set in
concrete and have a circular metal cover which fits over the
top. These manholes are only suitable for shallow domestic
drainage work.

The internal dimensions of inspection chambers increase as
the depth increases. The smallest size for brick and concrete

110

types less than 0.6 m deep is 0.6 × 0.45 m, and for circular concrete types, minimum height 0.375 m and diameter 0.9 m. The largest size for an inspection chamber over 4.5 m deep is 1.35 × 1.13 m for rectangular types and 1.8 m diameter for circular types. Generally, chambers over 2 m depth have a reducer slab or a taper section at the 2 m level. An access shaft is then built on top of this up to ground level.

To gain access to an inspection chamber over 7.5 m deep, step irons are built into one of the walls or cast into the concrete units. Step irons are galvanised malleable iron, having a tread which is also a hand hold and two legs which are built or cast into the wall. The top step should be not more than 0.45 m below the top of the metal access cover and the bottom step not more than 0.3 m above the benching. Steps are fitted at vertical intervals of 0.3 m and are staggered 0.3 m centre to centre.

The spaces between the channels, the branch bends and the walls are filled with concrete which is built up above the pipe and finished with a 1:1 cement mortar trowelled smooth – this is known as the *benching*. This benching ensures a through flow of water, prevents leakage in the manhole base and provides a foothold when a man is carrying out an inspection. The benching is usually formed in the base after the walls are constructed, but can be precast with a base slab which incorporates the channel. These precast base units in standard layouts are laid on the concrete base and are generally available for use with precast concrete inspection chambers, instead of the standard unit with cut-outs to accommodate pipes and branches.

How is a drain connected into an existing sewer?

Firstly, a hole is cut into the sewer pipe where the connection is to be made. For obvious reasons, this should be cut into the top half of the sewer pipe after first making sure of the level of flow in the sewer by inspecting at the nearest manhole. The hole can be cut by using a hammer and chisel, making a small hole in the centre of the proposed hole and gradually cutting away until the required diameter is reached. Alternatively, a heavy duty power drill may be used to make a series of holes round the circumference of the required hole and the opening finished off with hammer and chisel. A special fitting known as a *saddle* is then bedded into the opening with cement and sand mortar.

The saddle consists of a salt glazed clayware socket on a short length of pipe with a shoulder or flange shaped to fit over the outside diameter curve of the main sewer. The saddles are manufactured with various shaped flanges to fit different diameters of sewer and sockets to suit 100 and 150 mm drains. The pipe projects from the flange by about 25 mm and fits into the hole cut in the sewer.

Because the saddle fits at an angle, the hole cut in the sewer pipe is elliptical in shape. After the drain has been connected, the pipe and saddle are surrounded in concrete. Sewers are mainly the responsibility of the local authority and fitting saddles is carried out under the direct supervision of the authority.

Where the connection into the sewer is to involve the construction of an inspection chamber the following procedure is adopted. A hole is excavated large enough to construct the chamber and down to 150 mm below the underside of the sewer pipe. A concrete bed is laid to a depth half way up the sewer pipe. When the concrete has set and hardened, the pipe is cut using a power saw to form a channel, and branch bends are bedded in position. The inspection chamber is built up and benching completed in the normal way.

If the sewer can be diverted or stopped temporarily then it is possible to insert a junction pipe into the sewer pipe line. At least three pipes have to be cut out of the pipe line and two new pipes and a junction pushed into line and jointed. Alternatively, a double socket pipe can be used to join up to an existing spigot end, or if the existing socket is damaged.

How is a drain set out and levelled?
The position of a drain can be marked out on the ground by means of wooden pegs indicating the centre line of the pipe. The position of inspection chambers is simply marked by four wooden pegs indicating the corners of the chamber excavation.

The depth of trench excavation will always vary because the drain has a fall from chamber to chamber. To ensure that the base of the trench is excavated to the required gradient, a system of sight rails and traveller are used. The sight rails are erected next to, or across the trench line at inspection chamber positions, or at a change of direction or gradient on the main drain run. The rail consists of a horizontal wood rail fixed securely to one or two upright posts driven into the ground. Each rail is fixed at a suitable known level above the invert of the drain at that point. In conjunction with the sight rails, a traveller is used. The traveller is a T-shape, the long leg of the T being the same length as the depth from sight rail to invert level of pipe. When excavating the trench, the correct depth and gradient are maintained by using the traveller and sighting the top rail of the T and the sight rails and keeping them in line. The method is similar to the boning rod method of levelling described in Chapter 1.

Where the pipe is to be laid on a concrete bed, level pegs are driven into the base of the trench excavation showing the gradient of the concrete bed similar to foundation concrete. The pipe line is kept straight and true by means of a line stretched tightly from one inspection chamber to the next or from the junction to the end of a branch. The line is set on the centre line of the pipe run and along the top of the line of pipes or the top of the collar, if any.

What precautions can be taken to ensure safety while laying drain pipes in trenches?

Shallow drain trenches in firm soils up to 1 mm deep generally do not need support unless the trench is alongside a road, where traffic vibration may cause the collapse of the trench sides.

When trenches are dug for pipe laying the stability of the sides is a very important safety factor. Generally the soil is classified as *firm, moderately firm* and *loose*. Trenches in loose or waterlogged soil require close boarded timbering to support the sides at all times to enable the trench to be dug at all. Trenches in moderately firm soil may require open timbering, but this will depend on the depth and the weather conditions. In most cases trenches up to 1 m deep in this type of soil do not need timbering to support the sides unless weather conditions are bad or the trench is alongside a road. Generally the decision on whether to support the sides of a trench rests with the foreman on site, who is expected to be a competent experienced person. The Building Safety Regulations state that timbering should be carried out to prevent danger to persons working in the trench from a fall of earth from the side of the trench.

Open trench timbering consists of vertical members known as *poling boards* against the trench sides, supported by horizontal timbers called *walings,* with struts bracing across the trench. The spacing of the poling boards and the struts depends on the state of the ground. The spacing of walings depends on the depth of the trench. Close boarded timbering consists of poling boards which are side by side with walings and struts at very close intervals. Poling boards are generally about 200 × 50 mm, walings 100 × 150 mm, and struts 100 × 100 mm. Pairs of folding wedges are used with wood struts to ensure the tightness of the timbering. Alternatively telescopic steel props can be used instead of wood struts and wedges. Steel trench sheets which are interlocking can be used instead of poling boards, but are mainly used for deep excavations.

Where a trench is over 2 m deep a fence or safety barrier must be erected alongside to prevent persons falling into it. If the trench is alongside a path, a barrier and lighting must be provided to protect the general public. Materials must not be placed on the edge of a trench which may place undue stress on the trench sides at that point. Plant and equipment must not be used so close to an excavation that the sides are endangered. If concrete is to be tipped into the trench, an adequate stop must be provided to prevent the barrows or dumper trucks being used from rolling forward into the trench when tipping. Regulations stipulate that a weekly inspection be carried out of trenches and supports by a competent person and a report entered in a register.

12

PAVINGS AND KERBS

What are the tools generally used by the pavior?

Cutting and dressing tools. Hand cutting of kerbs, flags and paving materials is done with a 2 kg mash hammer and chisel, but for dressing stone and precast units a pitching tool is used instead of a chisel. The pitcher is a cold chisel with a broad flat end ground at an angle, similar in appearance to a bolster chisel but smaller and much thicker. The bolster chisel is used for cutting across the face of a stone or for cutting flags. Portable power driven saws are extensively used for cutting all types of pavior's materials.

Setting out and laying tools. Many of the tools used in paving are similar to the tools used by bricklayers described in Chapter 1. The square, lines, spirit levels, folding rules and trowels are all as used by the other trowel trades. In addition the maul, which is a rubber or wooden mallet, is used to consolidate or beat down precast units or stones. A laying pick is a tool similar to the brick hammer described in Chapter 1, but has a longer, straighter chisel end and is used as a general hammer and also as a lever to move units into position when laying.

What are the standard types and dimensions of kerbs, channels and flagstones?

Kerbs and channels are used to confine the edge of the carriageway and prevent flexible road surfacing being spread out by heavy traffic. The kerb forms a hard upstand which strengthens the edge of the footway or paving and acts as a guide line for drives, showing the edge of the road clearly. The kerb and channel also ensure the steady flow of surface water to the gullies.

Precast concrete is the main material used for kerbs, but natural stone is sometimes used in areas where it is necessary to match existing work. Kerbs and channels are manufactured in standard lengths of 900 mm and in a variety of sections. The rectangular and half-battered types are used where the foot-path is next to the road. The splayed type are used mainly where a grass verge exists.

All types of kerbs and channels are manufactured in a number of standard radii, both internal and external radius, ranging from 900 mm to 12 m. Edging kerbs are used for footways and paved areas for pedestrian traffic only. Flag-stones are used to pave footways and pedestrian walkways. They may be precast concrete or natural stone. The standard dimensions are:

900 × 600 mm × 50 or 67 mm thick
750 × 600 mm
600 × 600 mm
450 × 600 mm

In addition, paving is manufactured in a variety of sizes which, when bonded together, give a random pattern.

How are kerbs and channels laid?

The foundation trench for kerbs and channels is excavated or formed in the base course material of the carriageway. The concrete is compacted and levelled to level pegs in the trench similar to foundations for buildings. A semi-dry 1:3 cement and mortar is used to bed the kerbs and channels on to the foundation and the kerbs are aligned by means of a straight-edge and a line. On a long length of kerbing it is good practice to look along the kerb line at intervals of about 20 m to check the straightness of the kerbing. In dry weather the semi-dry mortar is damped after setting by spraying the kerb line with water. To stabilise the kerb, a concrete haunching is placed behind, up to 75 mm from the top. Boning rods as described in Chapter 1 are often used to level kerbs and channels.

How are paving flags laid?

Flags of natural stone or concrete have to be laid on a prepared bed of consolidated and rolled hardcore, blinded on top with a layer of coarse sand or similar material (*Figure 55*). The flags may be bedded directly on sand which has been carefully screeded to the correct level, or on to five spots of lime mortar on the sand bed.

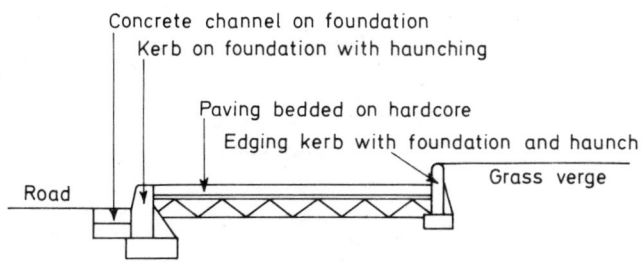

Figure 55. Typical footpath section

A flag lifter is a device which grips the edge of a flag and enables the flag to be lowered on to the bed in a horizontal position without trapping the hands. The advantage of this is that only a tap down with the maul is necessary to ensure a level finished job. The paving flags will settle down and become uneven and cracked if the base has not been prepared and consolidated adequately. The joints in paving flags are pointed in with cement mortar applied by pointing trowel or a semi-dry mix 1:3 is brushed into the joints. Coloured and decorative pavings are usually hand pointed and plain concrete flags brush jointed.

117

What standard bonds are used for paving?

Brick pavings are laid in a number of standard bonds, some of which are similar to brick walling bonds. Worcester bond, either flat or on edge, is a stretcher bond arrangement. Flemish bond laid flat is the same as 1½ B Flemish bond (*Figure 56*).

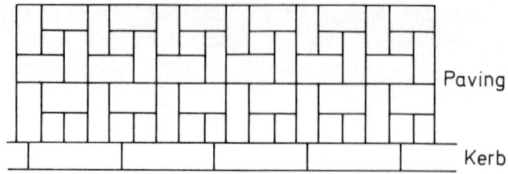

Figure 56. Flemish bond paving

Basket weave bonds laid flat or on edge are also used in brickwork. Norfolk bond and court bond are bonds specifically for paving.

How is brick paving constructed?

Standard bricks may be used for paving, provided they are a very hard and durable type, but specially made paving bricks are used in most cases. Paving bricks are the same size as standard bricks and may be laid on edge or flat. They are either smooth finish or have a pattern of diagonal grooves which give a non-slip surface in wet conditions.

The brickpaving is laid either on a sand bed or a mortar bed, on consolidated hardcore base. If the paving is to carry vehicles then the bricks may be bedded on to a concrete base. The joints between the bricks can be grouted with cement mortar 1:2, care being taken to ensure that the joints are filled and that no air is trapped. Bitumen joint sealing compounds are used as a joint filler for movement joints in large areas of paving.

INDEX